Practical Systems
Analysis

BCS Practitioner Series

Series Editor: Ray Welland

Practical Systems Analysis

A guide for users, managers and analysts

Roger Hipperson

Prentice Hall

New York London Toronto Sydney Tokyo Singapore

First published 1992 by
Prentice Hall International (UK) Ltd
Campus 400, Maylands Avenue
Hemel Hempstead
Hertfordshire, HP2 7EZ
A division of
Simon & Schuster International Group

Typeset in 10/12 pt Times Roman
by MHL Typesetting Ltd, Coventry

Printed and bound in Great Britain by
Dotesios Ltd, Trowbridge, Wiltshire

Library of Congress Cataloging-in-Publication Data

Hipperson, Roger.
   Practical Systems Analysis: A guide for users, managers and analysts
    / Roger Hipperson.
     p. cm. — (BCS practitioner series)
   Includes bibliographical references and index.
   ISBN 0-13-689688-X (pbk)
    1. System analysis.   I. Title.   II. Series.
T57.6.H56   1992
658.4'032—dc20

                                             92-15305
                                               CIP

British Library Cataloguing in Publication Data

TP

A catalogue record for this book is available from
the British Library

ISBN 0-13-689688-X

1 2 3 4 5 96 95 94 93 92

Contents

Editorial preface

The aim of the BCS Practitioner Series is to produce books which are relevant for practising computer professionals across the whole spectrum of Information Technology activities. We want to encourage practitioners to share their practical experience of methods and applications with fellow professionals. We also seek to disseminate information in a form which is suitable for the practitioner who often has only limited time to read widely within new subject areas or to assimilate research findings.

The role of the BCS is to provide advice on the suitability of books for the series, via the Editorial Panel, and to provide a pool of potential authors upon which we can draw. Our objective is that this series will reinforce the drive within the BCS to increase professional standards in IT. The other partner in this venture, Prentice Hall, provides the publishing expertise and international marketing capabilities of a leading publisher in the computing field.

The response when we set up the series was extremely encouraging. However, the success of the series depends on there being practitioners who want to learn, as well as those who feel they have something to offer. The series is under continual development and we are always looking for ideas for new topics and feedback on how further to improve the usefulness of the series. If you are interested in writing for the series then please contact us.

Roger Hipperson has extensive experience in training users and managers, as opposed to IT specialists. This is reflected in the pragmatic nature of this book and the viewpoint which is adopted. It is not a prescription for systems analysis, but a practical guide to the issues, supported by an extensive case study to illustrate the application of techniques described in the book.

Ray Welland
Computing Science Department, University of Glasgow

Editorial Panel Members
Frank Bott (UCW, Aberystwyth), John Harrison (BAe Sema), Nic Holt (ICL), Trevor King (Praxis Systems Plc), Tom Lake (GLOSSA), Kathy Spurr (Analysis and Design Consultants), Mario Wolczko (University of Manchester).

Preface

The author has followed the standard career path for an Information Technology (IT) consultant. I started by programming in low-level languages on a mainframe, progressing through a series of higher-level languages (such as Fortran and Algol), and playing with the so-called fourth generation languages. With this 'doing' experience behind me, I then moved across to the 'other side' and designed programming support tools, being there and thereabouts in the development of CASE (computer-aided software engineering), iCASE (integrated computer-aided software engineering) and IPSE (integrated project support environments). Until, at long last, I jumped the chasm and became an analyst.

An analyst is someone who specifies *what* has to be done as opposed to a designer who specifies how it will be done and the programmer who implements the design. Crossing the chasm has not been easy. Programmers and designers (after all, designers are usually programmers with a few years of experience in programming) live in a world of their own. They are given a document which they turn into a working program on a specified computer system. The programmer's day is spent trying to 'defeat' the compiler and stopping the persistent 'Error 7394 — Overflow' run-time error appearing on the screen. New techniques such as 'object orientated programming' and 'graphical user interfaces' must be tried. Concepts of sharing data and/or programs must be avoided as this will involve 'boring' discussions with other programming teams; and anyway, if the code is 'Not Invented Here' it cannot be worth using.

The analyst has to leave the programmer's worktop and suddenly acquire many interpersonal skills never before called upon. No longer are the analyst's ideas the most important: the analyst must interrelate with managers and users — two breeds of personnel which, as a programmer, were shunned — 'If it wasn't for users this program would be brilliant'.

There are many books on how to design and program, and a few books on performing analysis. Nearly all concentrate on the prescriptive techniques, taken from some (commercial) method, which the analyst, designer and/or programmer must go through in order to produce a working system. Few mention the analyst skills that are required to obtain the necessary information from users and managers.

This book describes the process of producing a computer system from the viewpoint of the managers and users acting in partnership with the analysts. Very few prescriptive techniques are described, more a set of basics which the system development team will be performing in some way, and how users and managers can help. Hence the book is primarily aimed at managers and users who wish to understand the basic building blocks used by the analysts; I strongly believe that all users and managers should be so inclined. The book will also be of use to analysts as an introduction to basic techniques and on how to interact with the 'other side'.

This book is based on my twenty years experience moving 'through the ranks'. Recently, I have been training users and managers in a number of large British organisations on the role of the analyst, and how the analyst can help in producing the system that the users and managers require and can afford. In performing this work I have been encouraged to record what I have been teaching; this book is the end result. I hope that the book's conversational, maybe even note-like, style will make a refreshing change from the normal textbook, and that you will find enough time to complete the book and so find out something (more) about computer system analysis and design.

The target reader

Who do I think you, the reader, is? Clearly, the title of the book implies that I am expecting users and managers to read the book; however, you may well be a member of one or more of the following groups:

1. A day-to-day user of a computer system that you feel would work a whole lot better if you had had some say in its design.
2. A day-to-day user of a computer system who cannot (for the life of you) understand why a certain facility was not implemented.
3. A manager of a group of day-to-day users (for example, the manager of some business centre).
4. A business manager who, although not a computer user, wishes to obtain timely reports on various aspects of the business.
5. A computer designer and/or programmer who would like to know what goes on before the design and implementation phases take place.
6. An analyst who would like an overview of the general techniques, perhaps before going on a specific methodology course, such as an SSADM course.
7. An analyst who would like a refresher course before setting out on an analysis exercise.

In fact, I expect that you may be anyone who is involved in the use, specification and/or building of computer systems.

Structure of the book

The major objective of this book is to explain the basic techniques used in analysis in such a way as to provide a complete description for users and managers, and a useful overview/introduction for analysts. In order to put the analysis techniques into perspective, the book places the techniques into the context of all of the work being performed on a project, including moving from analysis to design and then into implementation. To reinforce the descriptions of the general techniques there is a comprehensive case study. This case study moves from the start of a project right through to testing the final code.

The book reflects this mixture of context setting, descriptive sections and case study work as follows:

Part 1   A general overview of why one does analysis, and where analysis fits into the project life-cycle.

Part 2   Describes the general techniques used in analysis, including interviewing, presenting results, Business Analysis and producing Activity and Data Models. The first chapter explains the required interaction between users/managers and analysts.

Part 3   Introduces the case study and shows how the analysis phases are performed using the techniques described in Part 2.

Part 4   Reverts to the technique description mode. It describes the techniques used by the analyst in order to move from the analysis phase into the design and implementation phases. This part has been placed here because I feel that some understanding of how the analysis techniques are used is required before one can fully appreciate the techniques described in this part.

Part 5   Shows how the design and implementation plans are performed for one particular area of the business, including testing the code.

Acknowledgements

I would like to thank the reviewers of the initial drafts of this book for the many constructive criticisms. All of the comments were carefully considered, and most used to rework the text. I also thank the series editor, Ray Welland, for his help, together with Mike Cash and Viki Williams of Prentice Hall.

Grateful acknowledgement is made for permission to reprint quotations from the following books:

> Tom Peters, *Thriving on Chaos.* © Excel/A California Limited Partnership 1987. Reprinted by permission of Random House, Inc., 201 East 50th Street, New York, N.Y. 10022.

> Robert H. Waterman, Jr, *The Renewal Factor.* © Robert H. Waterman, Jr 1987. Reprinted by permission of Bantam Books, 61−63 Uxbridge Road, Ealing, London W5 5SA.

> Edward de Bono, *I am Right — You are Wrong.* © Mica Management Resources, Inc, 1990, p. 291. Reproduced by permission of Edward de Bono and Penguin Books Ltd, 27 Wrights Lane, London W8 5TZ.

In addition, I thank all those people I have worked for over the years, and so have put me into the position to be able to write this book, together with those, including my wife and children, who encouraged me to finish it.

Furthermore, I thank you, the reader, for reading this far and hope that you are encouraged to move on and read the rest of the book: after all, 'Nothing ventured, nothing gained'.

Roger Hipperson
April 1992

Part 1

Introduction

1 Overview: on the need for analysis

1.1 The reasons for doing modelling

One is quite often asked, Why do we need to do analysis? A simple and direct answer is:

> In order to agree a view of the business which everyone in the business can understand and agree to.

It is important that we understand the business first and then, and only then, build computer systems which match that understanding. Otherwise we will carry on, as in the past, building computer systems that do not match the business needs and, worse still, do not interact with other systems. We end up with a lot of computer systems each doing a particular task for a particular group. Quite often these systems cannot share data; and when they have to, one has to build yet another system in order to interface between them.

I once asked an analysis team I was about to join how the current system under investigation fitted into the rest of the world. The explanation went something like this. 'Well,' the analyst started, producing a diagram with a multitude of boxes on it (see Figure 1.1 for a simplified representation of it), 'the system talks directly, via magnetic tape, to system B. Systems C, D, E, F gather data from the system, but, because the data structures are different, we had to build a special interface system. Once a month the system under consideration produces a set of reports which are then sent to the other systems shown here,' the analyst continued, waving his hands over the diagram, 'and operators on these other systems type the data from the reports into their systems.'

Clearly, there was room for error (an understatement) in the existing system and there was an urgent need to analyse the current business needs and, although we were not able to solve today's problems overnight, to make sure that the organisation did not get into the same mess again and that the current mess did not get any worse.

Quite often a system designed from scratch without referencing other systems and much work is repeated, including the failures. Not only do we need to build systems that match an understanding of the business, we must also ensure that the agreed understanding is owned by everyone in the business. That is, users,

Figure 1.1
A hotch-potch of non-interacting systems.

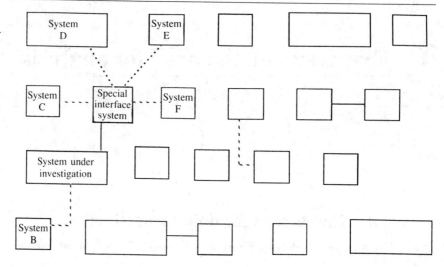

Figure 1.2
Commitment is required from everyone.

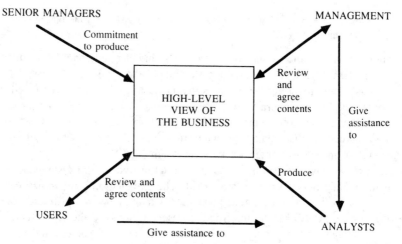

managers and analysts must be involved in the development of the understanding. Figure 1.2 shows how commitment is required throughout the development unit.

We start with senior management making a commitment to produce a high-level view of the business. Then management and users give assistance to analysts in order to produce the first pass view of the business. This view is then reviewed by the partnership until everyone has agreed its contents.

With this level of common understanding, the analysts and implementers can confidently move forward and produce systems suitable for the business and which offer no surprises to the users. Furthermore, the partnership can decide on the level of quality required for the system. A PC-based system to last six months will most probably need less attention than a multi-terminal distributed system with mainframes, workstations and PCs that has to be in service for the next ten

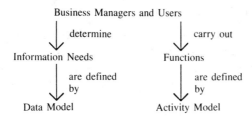

Figure 1.3
The pivotal Data and
Activity Models.

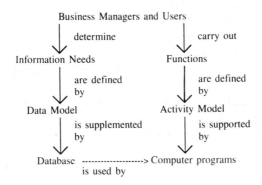

Figure 1.4
Information Systems
consist of a database
and associated
computer programs.

years or so. But be careful: what starts out as a short-term solution quite often becomes the foundation for a long-term solution, and so one must be wary of undercutting quality.

The theme is a partnership among the analysts, the managers and the users. The managers and users determine what their Information Needs are, and what Functions they carry out. Quite often this information has to be teased out by the analysts. This can be done in a variety of ways, including reading articles written by the managers and users, interviewing the managers and users, and presenting the structured information back to them.

The analyst structures the Information Needs and Functions into two models: the Data Model and the Activity Model (Figure 1.3). These first models are extremely important and become pivotal to the whole system implementation. The models must be at a sufficiently high level that the managers and users can understand them, yet detailed enough that analysts can further decompose them, adding more detail before passing them on to the designers. The models allow the complete design of the system to flow from the managers' and users' view smoothly into the design and implementation phases (Figure 1.4).

The pivotal Data and Activity Models are refined, then, until they are detailed enough for the implementers to produce a set of suitable data structures to hold the required information (database), and a set of computer programs and manual procedures to support the required functions. These together become the required Information System.

The data modelling exercise in the Analysis Phase is there to provide a common understanding. The Data Model can be used for at least the following:

1. The analyst can replay the interview to the original business manager, plus others, with a structured view. Misunderstandings can be readily identified (although not necessarily resolved).
2. The analyst, when drawing up the model, will identify areas in which there is doubt, missing information, or whatever. Again, questions can be raised to clarify the problems.
3. The analyst can compare the information given by one business manager with another, and identify problems such as:
    *   similar naming but different meaning;
    *   information and tasks for which no-one claims responsibility;
    *   duplicate sources of information, tasks and responsibility.
4. The analyst can present the model to his or her peer group for structured comment.
5. The analyst can present the model to other groups within the business.
6. The analyst can start involving the designers in ensuring that the system can be built and will achieve the requirements — and to start addressing technology options.
7. The designers take the agreed model as the basis for design.

Once we have a sound understanding of one area of the business we can share the understanding with other parts of the organisation. The analysts within a specific area of the business that requires a computer system should first trawl for existing models and see what can be shared. This will bring a number of immediate benefits:

1. The partnership in the new project need not start out with a blank sheet of paper, there being a number of existing models.
2. The different teams will interact with each other and an understanding of the data sharing can be obtained.
3. The analyst will not embark on an ego trip (producing yet another view of the business that no-one understands).

Tom Peters calls all of this 'Creative Swiping':

> Put NIH [Not Invented Here] behind you — and learn to copy (with unique adaption/enhancement) from the best! Do so by aggressively seeking out the knowledge of competitors . . .
> Become a 'learning organization'. Shuck your arrogance — 'if it isn't our idea, it can't be that good' — and become a determined copycat/adapter/enhancer.
>
> Tom Peters, *Thriving on Chaos*

Sharing the models will allow an organisation to produce a set of 'standard' models and/or a business-wide, 'corporate' model. Such models allow new projects to take the relevant parts of the model as a basis for their specific model and/or allow projects to check the completeness and consistency of their models. This is shown in Figure 1.5.

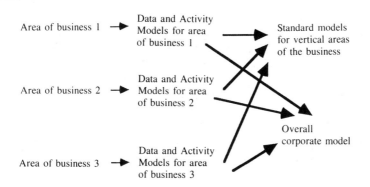

Figure 1.5
Producing standard
models and an overall
'corporate' model.

1.2   A simplified view of the main stages of modelling

We start with the determination of the objectives and targets for the whole business and split this into smaller business areas. Then information needs are determined and structured into data models. In parallel, the scope of the system is documented using a Context Diagram. The Functions of the system are documented using a Functional Decomposition Diagram and further refined using Data Flow Diagrams. Lastly, these views are reconciled with, for example, the data definitions given on the Data Flow Diagrams matched with the attributes defined on the Data Model. Much use is made of matrices to perform this task.

The first task is to determine the information needs required to support the business objectives. Too often we jump in and start coding the computer programs (by typing 'begin' or whatever) without considering the needs of the business. We have to stand back and ask whether we are doing the right things right. This is even more important if the system is to last several years and the information held on it is to be shared with others.

The aim, objective even, is to determine the objectives of the business; and then for each objective, to determine how to measure how the objective is fairing. Once the measures have been determined, the targets for the measures, which will decide whether the objective has been met or not, are stated. The targets are very important, as they ensure that the objective is sensible and that its success or failure can be measured and judged.

Objectives are an important starting point and it is essential to ensure that:

1.   The objectives are right.
2.   The objectives are the ones that concern the business.
3.   The objectives are achievable (may even be being achieved now).

1.3   The system development life-cycle

There is a famous set of pictures describing the average 1960s life-cycle based on building a swing. The following is a suitable variant.

Figure 1.6
User's rosy view of a
swing.

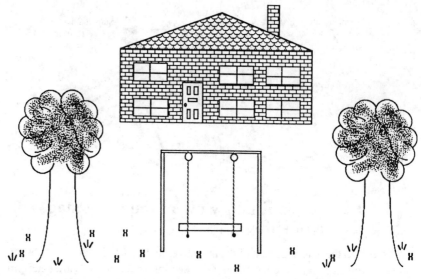

Figure 1.6 shows what the user wants: a pleasant environment with trees, flowers and a nice house. In all of this he wishes to have a swing. But no-one knows what a swing is and so it has to be described.

The analyst is a hard-nosed individual and the IT manager wishes to cut costs, and so the analyst perceives the need as shown in Figure 1.7.

Figure 1.7
Analyst's view of the
swing.

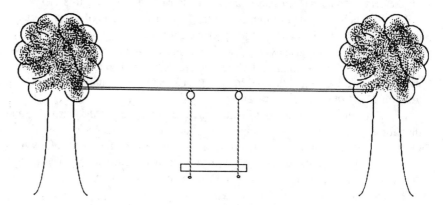

Hence, although the basic concept of the swing has been carried forward, many of the niceties have gone. Note that maximum re-use of existing materials.

The implementer scales the thing down to an absolute minimum and, look you lucky users, we can implement two for the price of one (Figure 1.8).

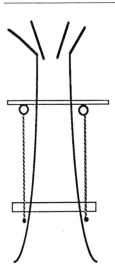
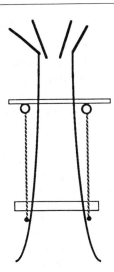

Figure 1.8
Implementer's view.

By the way, what did you want the two planks of wood hanging from some string for?

QUESTION: Why do you think this situation occurred? (Rhetorical)

High-level and fourth generation languages (4GLs) meant that more code could be produced faster, leading to even less time talking to the users. There was a myth (still around?) that said that if we can build a system quickly we could then alter it quickly to meet the real needs of the user. In fact, we must get the basic structure correct, that is, data definitions and function distribution. So it was realised that techniques were required in order to capture the user's requirements and then to ensure that the requirement was passed smoothly through the development life-cycle, from concept to the grave.

Many techniques blossomed, basically split into 1960s technology based on the Yourdon style of modelling and the 1980s technology based on the James Martin/CACI Information Engineering style of modelling. Many companies now realise that none of these is a total solution, and company-specific techniques are being generated, usually made up from the existing ones. These techniques concern not just program design but also control of the project and of the overall life-cycle. Life-cycle control techniques tend to be called iCASE (integrated computer-aided software engineering) and IPSE (integrated project support environment). Unfortunately, iCASE and IPSE have not, as yet, succeeded for a variety of reasons — mainly cost and the lack of metrics to prove that the money is worth spending. Project control techniques used by the UK government have been dominated by PROMPT and now PRINCE, both CCTA approved techniques.

There are automated support tools for many of the techniques — usually called CASE tools. No one tool is adequate enough to do the whole life-cycle, despite what the suppliers claim, and most large organisations utilise a wide range of tools.

The basic life-cycle that all projects must go through is:

1. Understand the environment in which the system must work.
2. Ensure that the users know what they want, the system designers know what the users want, and that everyone knows what they are actually going to get.
3. Detail the previous understanding to ensure that all of the logic of the system is understood, complete and consistent.
4. Map the *logical* system onto how the system is going to be implemented — the *physical* system.
5. Implement the physical system using techniques suitable for the computers being used.
6. Test, gain acceptance and review the system's use.

It is essential that each project has a document (sometimes called a Quality Plan) detailing the techniques being used. This document should also make the notation used clear enough for senior managers and users to read, and so to understand the models being produced. Otherwise it has been demonstrated time and again that the 'business' will not get the system that is required.

Lastly, a true story about a well-known firm's sponge cake mixes. Those of a certain age will remember these packaged sponge mixes which you make up in some way, bake, and then put lots of raspberry jam inside and lots of icing sugar on top. Some time ago, this manufacturer launched a miracle just-put-it-in-a-tin-and-bake-it mix which, despite being heavily advertised, failed to take off in the marketplace. The firm performed a market survey which identified the problem, and replaced the wonder mix with a 'simpler' one that required the consumer to spend time:

    adding an egg
    adding milk and water
    mixing it all together

What do you think the main result of the market survey was? It was that *the user wanted to be a significant part of the process.*

This book concentrates on techniques which will allow the user (and manager) to be a significant part of the process in specifying an application system.

Part 2

Analysis: general techniques

2 Overview of what users and managers should expect from the analyst

2.1 Us and them

This book concentrates on the basic techniques used by the analyst in order to interact with the users and managers within an organisation. Users and managers do not need fully to understand how to apply these techniques; rather, they should understand what techniques are being used and how the users must interact with the analysts using the techniques.

As with all professions, the computer industry has built its own wall of special words to describe the techniques being used. These, it is claimed, help the practitioners to interact with each other. Far too often, even the practitioners get confused, especially when new buzzwords come along (get someone to explain the differences between CASE, iCASE and IPSE (all meaning a set of tools to help one build a computer system), let alone the trendy (in 1991) 'object orientated' this and that). The idea, of course, is to create an impregnable wall between the highly paid computer staff and the business 'lay' person.

This has got to end: the Berlin Wall has gone — let us get rid of the Information Technology wall. Otherwise users will *not* get the system that they require to perform the business functions that are needed in order to manipulate the business information.

Whatever the justification for such Chinese walls, it is important that the *analyst* (not the user) crosses the wall and presents the results of analysis to the user *in a manner that the user can understand*. This is the paramount concern when interaction between users and analysts takes place. If this means that the analyst must spend time transforming the structured results into a user-understandable format, then so be it.

However, the user can help. One of the main problems of Information Technology projects in the past has been that the projects generate a set of disjoint deliverables. Quite often, even today, the initial documents describing the high-level requirements are produced, misunderstood by all, and then placed on the library shelf. The design and implementation of the system is then performed based on the analysts' and designers' view of the proceedings which took place in order to produce the initial requirements documents. Even in an ideal world, one would be extremely lucky if the completed system matched the needs of the users. The project team can then spend the next ten years 'enhancing' the system as the criticisms come pouring in.

Most large companies do not perform in an 'ideal' world. Quite often the initial feasibility study is performed by outside consultants, more or less on their own. Some companies have tried to set up a joint team with a token internal analyst sitting somewhere in the team, but always led by the outside consultants. This situation can be worse than using internal staff only. The external consultants will be experienced analysts (otherwise the organisation wouldn't have employed them) and will want to get on with the job, providing a detailed analysis of the problems and solutions. The external consultants will 'own' these deliverables, and only they will have a complete and overall understanding of the results — no matter how well they write the final deliverables. This makes it difficult for the internal team to pick up the deliverables later for the next stage of the project.

Sometimes the friction between external consultants and internal staff may be so great that their use in the way just described is simply a waste of money. However, the use of external consultants is essential for most companies and projects during the initial phases of a project. They bring to the table a fresh view of the business area and 'new blood' to the team. In order, however, to overcome the inherent problems, the consultants must be 'seconded' onto a project team and so not be a major part of it, nor lead it. The consultants should not be responsible for publishing the report: the report must be owned by the internal project team and the style must be theirs, and the report should carry the project's logo on it. This sentiment is true even if the 'external' consultant is a consultant brought in from another part of the business.

I advocate the use of 'skills transfer' whereby the external consultant joins the team and helps the team members to acquire the necessary skills. This method of working, further detailed below, may slow down the initial work, but ensures that the project team understand the deliverables being produced and so feel that they *own* the results.

2.2   How to get the users involved

Users and analysts are very wary of each other. The users have the business knowledge and, for better or worse, they are making the business work. They seldom have the time or specific IT-related expertise to translate their business knowledge into support systems — automated or otherwise. So the users pay 'loads of money' to IT staff to perform this transformation. However, the IT staff do not, in general, have much direct experience of the business — just the automated systems that may or may not adequately support a part of the business.

Hence the two parties must talk to each other, both at formal meetings (see Chapter 3) and at informal meetings, say in the canteen. Each party must not be afraid to ask questions in order to gain a better understanding of the area of business and its support systems.

I remember quite vividly my first day in a particular organisation. First of all

I read through some of the written background material, listened intently to an overview of the current automated support systems, and deduced a number of business needs. Then I asked a large number of basic clarification questions, most of which were answered satisfactorily. I then asked some 'dumb' questions, questions which the receiver of the questions felt were a little silly as 'it was obvious what the answer is'. (I quote Tom Peters on the subject of dumb questions in Chapter 3.) As usual, some of these 'dumb' questions provoked a certain amount of merriment. However, in my egoless analyst role, I swallowed my pride, thought of the money, and carried on. I knew that I was obtaining a picture of the business system in existence, its problems, and the requirements for the new system.

One of the questions did not cause mirth, but instead a frown appeared on the IT staff's face. 'Well, we have always done it that way,' the voice said worriedly, 'but I agree that it needs to be questioned, as I do not know why we do it that way.' 'Good,' I thought, at least this question was not so dumb after all. 'So what do the users say?', I asked naively. 'The users!', exclaimed the analyst in horror. 'We haven't asked them.' 'Why not?', I asked. 'Because we don't want to show them that we don't understand their business; and besides, there's a lot of politics between us and them.' My jaw dropped and I re-read my contract with a view to tearing it up. But no, wait, I was there to spread the word. 'Come on', I arrogantly commanded, 'who should we talk to about this?' The next day we duly turned up at the relevant senior user's office. Going against the grain of skills transfer (the subject of the next section) I let myself be bullied into asking the question. In response the user looked at us and agreed that he did not know why the thing was done in the way it was. He would go and find out.

Eventually we found out that there was a sound business reason for doing the thing I initially questioned in my dumb manner. However, everyone in the chain now realised why it was being done and why there needed to be continuing support for it. The analyst was happy. The user felt happy, knowing that the analyst was taking due consideration of the business needs. This user and the analyst struck up a good working relationship for the rest of the project.

2.3   Skills transfer

It has been the general rule within many companies to employ consultants to perform a specific task under time and cost constraints. This normally means, with the agreement of the company, that the consultants perform the task on their own, usually to their own standards. This method of working has the benefit of obtaining some sort of deliverable/system in a timely manner, but has all of the attendant problems of:

1.   Company staff not being fully involved in the study, design or whatever.
2.   Company staff not feeling that they have the ownership of the deliverable.
3.   The deliverable being difficult to maintain.
4.   Company staff being unable to repeat the exercise at some later date, leading to a repeated cycle of requiring yet more external consultants.

These problems can also occur when an internal consultant is drafted in from another part of the organisation.

I believe very strongly that any organisation with a reasonably large Information Systems department, say greater than ten members, should ensure that its IT staff obtain the skills necessary to perform the many roles that they have to perform. This can be achieved in a number of ways, including:

1. Specific training on outside courses.
2. Specific training on organisation-specific courses.
3. Recruiting highly skilled staff.
4. Skills transfer.

The first of these is a standard technique but liable to a number of concerns. First of all, outside courses tend to teach the basic techniques for a particular method without regard to the particular circumstances of the attendee's organisation. How many people go on Data Modelling courses? Probably not very many — people go on an SSADM course or some other such specific method course where the method's prescriptive view of the technique is taught. Nevertheless, these courses are very useful, but only as a first pass in order to recognise and understand the techniques.

The second of these techniques is somewhat more expensive than the first to set up, but far more cost effective in the long run. The base material for such 'client-specific' courses can, and quite often is, from a standard public domain course and is, in some way, bought in. The course can then be tailored to match the organisation's techniques, and organisation-specific case studies generated.

Such client-specific courses can then be run to suit the particular project needs; for example, on-site or at a hotel, using project-specific CASE tools, and/or spread over a suitable period to match the project needs. There is no point in training someone too early, otherwise it will be forgotten before there is a chance to use it. The format of the training material can be customised as well. Some material I have worked on has been packaged in a number of different styles, each aimed at a different cross-section of the project members:

1. Straight lecturing with some in-line paper exercises.
2. Combination of lecturing and project-specific workshops. The workshops included the use of CASE tools, organisation-specific case studies, and some practical work based on the attendee's own project material.
3. Self-teach material aimed at project staff who would not be able, or would not be interested, in attending the other styles of presentation. Such material has to be an overview, and include many self-check activities. The material is aimed at senior IT staff, users and managers, and perhaps at novice IT staff who require a quick introduction before tackling the in-depth material.
4. Self-teach material covering the same ground as the lecturing material, aimed mainly at analysts who are likely to look up the details in project-related manuals (for example, the company standards) and who are likely to attend associated workshops.
5. Practical workshops to support the in-depth self-teach material.

The third technique — recruiting highly skilled staff — has a certain amount of merit, certainly all of that expertise is owned by the company (until the highly skilled staff leave). However, there is still likely to be a need for some retraining in order for the new staff member to become used to the specific techniques used in-house. There is a tendency in industry for senior managers to expect far too much out of newly recruited senior staff. They have to adjust to a new environment, one in which they hope to stay for some while — and this takes time. And they cost a lot of money.

To skills transfer, then. Here the company accepts that the project team does not have all the skills necessary to perform the tasks, and embarks on training them instead of employing new, highly skilled staff. Consultants are brought in who have the necessary skills and sit with the team, advising them how to perform the tasks. One consultant per phase of the project is my recommendation. At the Analysis Phase, employ a consultant with a wide range of analysis skills obtained in a wide range of companies. In the Design Phase, the suitable consultant may or may not be the same as that used for the Analysis Phase. Clearly, as one gets into implementation, one may or may not employ a group of contractors to perform in particular areas, although one may still require a specialist, say in the area of database tuning. Again, although I have emphasised the use of an external consultant — if only because they bring 'new blood' — the consultants used can be in-house staff.

The skills transfer consultant, then, is employed to *aid* the project team in performing some part of the project. The idea is to fuse the consultant's considerable analysis skills and experience to the project team's. The consultant will show by example to start with, but then slowly ensure that the team has the confidence to own the deliverables and understand the completeness and consistency (the status) of the deliverables.

The types of skill that need to be transferred at the Analysis Phase include:

1. The ability to plan what information is needed, and when, in order to perform the analysis, including the availability of key users and managers for interviewing.
2. The ability to ask 'dumb' questions and listen to the facts, then to follow up with the not-so-dumb questions.
3. The ability to step back and think of the overall business data and functions.
4. The patience to draw the relevant models in a form suitable for users, analysts and designers (this may mean three separate models).
5. The patience to write up the models with suitable narrative, business questions and examples.
6. The patience continually to question the structure of the models, and so to determine if they are correct, whether all of the necessary questions have been asked and answered satisfactorily.
7. The ability to document the model using the right tools at the right time.
8. The ability to present the models to peer groups, senior IT management, designers, users and business managers.
9. The ability to describe the pictorial models in a suitable English text,

remembering that a picture is *not* worth a thousand words, and that every picture needs the thousand words behind it before it can be adequately understood.

10.   The ability to model to the level required.
11.   The ability to produce consistent models in set times.

The types of skill required for other phases depend very much on the type of project in hand. It is impossible, therefore, to give a generic prescriptive list here.

2.4   Users and managers must not be shy

I was giving a talk to a group of senior managers on the need for a common understanding, when I was interrupted by one older, 'I'm retiring next year' participant who pointed out that he could not understand the complicated pictures being shown to him by the analysts.

'Well, what do you do about that?', I asked. 'Give up trying to understand. After all, I'm paying the IT people a lot of money to get the system right for me, so why should I worry? Besides, I don't wish to admit that I can't understand their modern, high-tech deliverables.' I screamed and carefully disengaged myself from the ceiling. 'Don't be shy,' I say, 'tell the analyst to simplify the deliverables.'

Analysts are good at blinding users and managers with their science. Remember that they were once programmers and designers, working in an environment where, it seems, you have to impress your peers. One had to have a better, 'more elegant' way of producing a piece of code than they did, and only you knew how to beat the compiler in such and such a way. Analysts, growing out of this environment, look for better and more impressive ways of producing the analysis deliverables, usually by making them more complex — 'my picture's better than yours because I've got 400 more boxes on mine and loads more lines'.

Tha analyst must outgrow such nonsense and produce deliverables which are suitable for the audience that is going to review them. If the reviewers cannot understand the deliverable being presented, it is not their fault — *it is the analyst's fault*. The user/manager must not be shy, the user/manager must tell the analyst to go away and re-present the material after it has been simplified. Of course, in some cases the user/manager may need to brush up on some of the relevant analysis skills, and the analyst should be prepared to present/lecture the relevant techniques if asked to do so.

3 On interviewing and presentations

3.1 Introduction

Interviewing and feeding back the information are the basic information-gathering and checking techniques. Interviewing is essentially one-to-one: there may be others present, but one analyst is trying to understand the needs of one user/ manager. Feedback Presentation is the technique for presenting what has been done for comment, both to peer groups and to managers and users. Rapid Interactive Analysis is a combined interviewing and feedback presentation technique used with a group of users and managers to gain information rapidly in an interview, feed back and discuss environment. These techniques are described in the following sections.

3.2 Interviewing

As discussed in Chapter 2, the interaction among the analyst, users and managers is vital if we are to understand the needs of the business.

The first formal point of contact is an interview. Interviews may be conducted during any stage of the project as the need arises to gather more information about the business or system. The bulk of the interviewing will be performed during the business analysis stage. There are many books on the art of interviewing. This section describes the basic techniques that I, and many others, use when interviewing. One must always remember to adapt the technique to the environment in which it is being used.

When I go into a new situation as an analyst, I introduce myself to some of the team, most usually through the project member who recruited me. I then ask someone to spend an hour or so explaining to me what the overall project/problem is about — that is, give me the background information. I mainly listen, only asking questions when I do not understand the meaning of a sentence. I will tend to ignore some of the business-specific terminology at this stage if I can make intelligent, even if incorrect, guesses. The idea is not to spoil the flow of information. Once the flow of information falters and/or becomes too detailed, and/or after about 30 minutes, I bring the session to a close, asking for any additional written background information. Information provided in brochures for the general public, such as company annual reports and in-house magazines, are excellent sources of information, as they tend to explain the business-specific terminology.

From this initial background information I can pick up a feel for the project and ensure that I have a list of nouns and verbs that describe the business area. I then generate a 'glossary of terms', a list of business functions and a list of the important data requirements.

A colleague and I performed some work for a large government organisation where the computer systems were dealing with differing payment systems throughout the country. We spent the first few weeks listening to the many mnemonics and put together a list of about one hundred terms. This we carried around with us to meetings and continually referred to it, adding more terms as necessary. The organisation's staff noticed this, and we soon received requests for copies of it. The moral of this tale is that even the long-standing staff of an organisation tend not to know what all of the terms used in business mean, and the point of performing analysis is to spread common understanding, even if producing a glossary of terms does not fit into the standard prescribed way of doing things.

The interview is the users' and managers' opportunity to ensure that the analyst has a good understanding of the business needs, which problems need addressing and what the business opportunities are. Hence the interview sessions are extremely important and all parties must ensure that there is adequate preparation for the session. This means that the interviewee must understand what the objectives of the interview are.

Although it is really up to the interviewer to prepare a list of questions and guide the interview, it is also important for the interviewee to:

1. Have a list of points that must be addressed.
2. Help prepare a suitable environment (mainly a suitably sized room) where there will be no interruptions — especially from telephone calls.
3. Be prepared to help the interviewer understand the answer by:
    *   being patient with repeated questions;
    *   being willing to rephrase the answer to help the understanding.
4. Be willing to consider other people's point of view in the sense that when answering a question, warn the interviewer, if necessary, that another user/manager may have a different view.

The point about being patient is important. In order to promote understanding, the interviewer has to ask many 'dumb' questions. Tom Peters, in *Thriving on Chaos*, states that:

> Mostly, it's the 'dumb', elementary questions, followed up by a dozen even more elementary questions, that yield the pay dirt. 'Why the heck does this form go there next?' and 'Who has to sign it?' are probably the two . . . vital questions [in many circumstances].

Two very important aspects of an interview are: (1) the preparation by both parties, and (2) the taking of notes by the interviewer (analyst). Both the interviewer and the interviewee should have a set of questions and/or points that need addressing during the meeting. The user/manager must ensure that the interviewer is taking copious notes, otherwise the valuable time the user/manager is putting into the interview is likely to be wasted. Here is Tom Peters again (from *Thriving on Chaos*):

> Engaged listening may be the principal mark of concern that one human being can evince for another, in any setting.

> The note-taking habit is a tip-off to whether or not engaged listening is taking place.

What do I mean by 'copious notes'? I certainly do not intend a verbatim statement of what is said. What is required is a list of all the key points, and especially the things that the user/manager expresses strong opinions about — his 'hot buttons'. For each of these key points the analyst needs to document their importance and the understanding of what has to be done to address the point. What the note-taking is doing is making sure that the interviewer has understood what has been said. It is amazing how difficult it is to write these notes; our brains assume too much when we listen to someone — 'unloading' our brains ensures that we really understand what was said. A good example is our general acceptance of mnemonics during a conversation, even though deep down we know that we do not know what it means, let alone what the letters stand for.

Making notes during the meeting certainly gives the listener a headache and may require the speaker to stop talking for a whole. However, the interviewer can be sure that he has, more or less, understood the point if it can be written down in note form. If there is enough doubt about the point, a question can be asked, such as, 'I just wish to clarify that I understand what you have said.' The user/manager can help the analyst by making sure that the important points are well understood, for example by asking the analyst to replay the last point. Again, I state that the most important aspect of an interview is to pass information so that a common understanding can be obtained. This will not happen if one worries about one's ego too much and shies away from asking 'dumb' questions.

Lastly, at the end of the interview make sure that the next steps are well defined, for example, a feedback presentation date. Remember that the analyst is attempting to build a system to suit the manager's/user's business needs: therefore, the analyst needs as much help as possible. One final aid may be a list of other users and/or managers who should be interviewed in order to obtain a complete view of the business needs.

3.3   Feedback Presentations

The information gathered by the analysts from reading background material and interviewing is put together in a structured manner. Structuring the information gives the analyst an opportunity to discover inconsistencies and to find areas which may not be as well understood as is necessary (that is, the existing information is incomplete).

The analyst should then present the structured view to a peer group. This will:

1. Help the analyst 'learn' how to present the material in a coherent manner.
2. Pick up obvious errors.
3. Help identify inconsistencies and incompleteness.

Then the analyst, perhaps with the analysis team, should present the material to a group of users and/or managers. The list of the main objectives of presenting the structured information back to users and managers is quite long, but here are four important ones:

1. To identify gaps in the analyst's knowledge.
2. To ensure that there is common understanding of the material, and if not, to obtain it.
3. To agree on the naming of parts of the structure: for example, ensuring that data items have an agreed name and meaning, and that activity names are meaningful.
4. To ensure ownership (remember the cake mix).

There are two formats for presenting the structured information back to users and managers:

1. A walkthrough of the structured material, clearing up misunderstandings, gaining common understanding and identifying, but not solving, problems/ areas of uncertainty. (These types of feedback presentation are often referred to as structured walkthroughs — this format I have called the Feedback Presentation.)
2. As (1), only with the addition of interactive sessions to solve the identified problems. (These are often called JRD or JAD sessions after IBM's Joint Requirements Design and Joint Application Design sessions — this format I have called Rapid Interactive Analysis.)

It is important that the presenter and audience clearly understand which format is being used. Presentations in format 1 should take no longer than two hours (the optimum is around an hour). Presentations in format 2 should be carefully scheduled with a structured agenda and an experienced IT orientated manager 'facilitating' the meeting. As there is likely to be much discussion before an agreed 'solution' can be found, such presentations are likely to take at least four hours, quite often the whole day.

In general, there are four groups involved in a Feedback Presentation:

> The Chairman
> The Note-taker
> The Presenter
> The Reviewers

The Chairman must be unbiased and should preferably have no direct involvement in the project area being presented. The Chairman must ensure that the Presenter is given the opportunity to present the deliverables, and then preside over a period of objective questioning. The Chairman must aid the Presenter in so far as if the

Presenter is unable to answer a question and/or the question starts a general discussion on possible solutions, then the Chairman shall call a halt to the topic, get the Note-taker to document the problem, and then move swiftly on to the next point. Feedback Presentations are an arena to present the work done and to hear the levels of concern from the Reviewers. The Presenter will have to solve any problems outside of the meeting; this is what keeps the time period down to a manageable hour or so.

The Chairman is also responsible for setting up the Feedback Presentation, distributing the necessary documentation in advance, chairing the presentation, and then ensuring the necessary post-meeting actions are carried out.

One area of these presentations that tends to go wrong is the version control of the deliverables being presented. The timescale from 'freezing' the deliverables for copying and distribution to the actual presentation may be some weeks. On a live project, some of these deliverables may be shown as incorrect and so updated. If the Presenter, as too often happens, presents an updated version of the deliverables that were reviewed before the meeting, the Reviewers are likely to be a little confused. Conversely, if a Reviewer fails to review the documents before the meeting, then it is likely that the Reviewer will not be able to participate fully in the review process.

The Chairman's job is to ensure that the time period between circulating the deliverables for review and the presentation is long enough for the documents to be reviewed and yet short enough so as not to affect the project timescale. The most important criterion is whether the Reviewers will have enough time to comprehend the deliverables and to formulate any questions.

The Note-taker is sometimes one of the Reviewers, but, in my opinion, this does not work too well — better that the Note-taker is one of the project team whence the Presenter comes. The key notes required are the list of errors found together with the names of the persons responsible for ensuring that each error is resolved. The final note will be the status of the deliverable, being one of:

1. Deliverables rejected (out of hand) and project recommendations for the future.
2. Deliverables accepted as they stand.
3. Deliverables having a number of errors, all of which seem to be resolvable via the list of errors and responsibilities detailed in the notes. The Reviewers will trust the Chairman to complete the signing off of the deliverable.
4. Deliverables having a number of errors, some of which are of a serious nature, and will require a further Feedback Presentation on this set of deliverables in order for them to be agreed. The Chairman is to notify the Reviewers when the deliverables are ready for another review.

The Presenter is one of the team who produced the deliverables. The Presenter must give a technical presentation on the material and answer technical questions about it. The Presenter should neither try to defend the work nor enter into discussions on better ways of doing things. The Presenter must first of all communicate the structured material, and then listen to the questions and criticisms, responding only if the questions are largely on uncertainties of understanding.

The Presenter must be *egoless*. The Presenter is part of a partnership and is trying to structure a common understanding. Similarly, the Reviewers should not be too harsh. The Feedback Presentation must not turn into a confrontation. Edward de Bono, in his book *I am Right — You are Wrong*, describes a 'new' way of thinking called 'water logic' as a contrast to the traditional way of thinking:

> In a conflict situation both sides are arguing that they are right. This they can show logically. Traditional thinking would seek to discover which party was really 'right'. Water logic would acknowledge that both parties were right but that each conclusion was based on a particular aspect of the situation, particular circumstances, and a particular point of view. (p. 291)

It would be up to the group, perhaps guided by the Chairman, to understand the different aspects of the situation, the particular circumstances and the particular points of view, and to help the Presenter and analysts to reconcile them.

There are two types of Reviewer. The first are members of the peer group of the Presenter, and these Reviewers are checking for technique errors and allowing the Presenter to 'rehearse' the presentation. The second type of Reviewer is the users and managers who are checking for the correctness of the deliverables with respect to the business.

Such Feedback Presentations are an essential part of producing complex deliverables and are especially important for Data and Activity Models. Such models are there to add to everyone's common understanding of the whole system. Feedback Presentations help to ensure this common understanding among the Reviewers and the analysts; the reviewed model(s) can then be distributed to others in the knowledge that the material is of a suitable quality to be understood by others.

Lastly, it must be made clear to everyone at a Feedback Presentation that it is a major opportunity to ensure a common understanding of the solutions being proposed. Everyone at the meeting should contribute, as being silent will be taken as acquiescence.

Remember that the analyst is likely to be wrong. The analyst structures what is available and the Feedback Presentation is the forum where the analyst is asking for agreement, guidance and commitment to a solution. In fact, we must be careful about this right or wrong confrontation, as the requirement is a mixture of common understanding and agreement with a good slice of experience thrown in. Those with much business experience will sometimes know the requirement without necessarily being able to articulate why this is the best approach. The job of the analyst is to ensure that everyone agrees with this 'subjective' view. It is as well to remember the chess player, though, as related by Robert H. Waterman in *The Renewal Factor*:

> British psychologists Peter Watson and William Hartson, the latter an international chess master, lend their version of 'away from goodness', by re-creating how two chess players analyze a particular move:

Weaker player: What's wrong with it?
Stronger player: It's not good.
Weaker player: Why not?
Stronger player: It's not the sort of move you play in this sort of position.

End of conversation

Analysts, just like the stronger chess player, have to have a good deal of information in their heads, gained from experience, in order to help them use intuition to decide if something is the 'right' way of doing something. This is why there are few books on practical analysis, yet many on prescriptive techniques.

3.4   Rapid Interactive Analysis

The previous techniques of interviewing and feeding back the structured results of the interview(s) are the key ones used by analysts and are unlikely ever to be completely superseded by other techniques (not until automatic brain scanning is perfected, anyway).

These techniques gather and confirm the correctness of a lot of information, even though the information may be disjoint or incomplete. Whilst talking with one user or manager, the analyst may obtain information that is widely different from the 'facts' obtained from another. Unless guided carefully, and this is not at all easy, users will tend to answer the direct questions with just the information required to satisfy the questions.

Rapid Interactive Analysis sessions are an opportunity to gain information and test its consistency and completeness at the time of gathering. The sessions are basically workshops in which the users and managers are the key participants, and the analysts are there to assist the users and managers to structure and understand their requirements. The original concept comes from IBM's JAD (Joint Application Design) sessions where end-user screen designs were 'thrashed' out. These were extended to JRD sessions for analysing the user requirements for a system.

The Rapid Interactive Session must be extremely interactive and dynamic. The 'unstructured', although guided, information from the users and managers is structured *during* the meeting and played back to the participants for comment. The key attributes of a Rapid Interactive Session are:

1. It is a concentrated workshop.
2. It is attended mainly by users and managers.
3. It is conducted by a skilled 'facilitator', who will structure the discussions.
4. It is documented by a skilled analyst who will present the discussions in a structured manner.

Thus, the Rapid Interactive Session helps to short-cut the tedious one-to-one interviews and individual review of deliverables. Unfortunately, such techniques may still be necessary, especially near the start of the project, but there is a need for users and managers to hear each other's points of view, and for the group of users and managers to come to some understanding and agreement. Listening to each other's points of view helps this process and allows the analyst and the decision-maker to make a more considered structured view of the requirements.

This leads to a number of significant benefits for performing a Rapid Interactive Session, namely:

1. Ownership of the system under discussion is spread throughout the organisation — and rests with the users and managers *(not with the analysts)*.
2. Everyone feels that they have had a significant part in the decision-making.
3. The quality and usability of the system is improved.
4. The overall design process is faster due to improved understanding of the problems and joint agreement of the solution.

Here is a list of the sorts of topic that would be discussed and agreed within a Rapid Interactive Analysis session for each of the project life-cycle stages:

1. During the Business Analysis Phase:
    *   Agree objectives and targets;
    *   Agree the things which will help and hinder the success of the identified objectives;
    *   Agree the scope of the project;
    *   Consider the impact of technology on the proposed system.
2. During the Activity and Data Analysis Phase:
    *   Agree and finalise the Data Model;
    *   Agree and finalise the Activity Model;
    *   Agree and finalise the ownership of data.
3. During the Design Phase:
    *   Agree and finalise screen designs;
    *   Agree and finalise report content and layout;
    *   Review any prototyping;
    *   Agree and finalise the overall logic of the activities.

It is unlikely that Rapid Interactive Sessions would be used during the Implementation Phase, as the users and managers need not be concerned with how the agreed design is actually implemented.

Rapid Interactive Sessions are far harder to organise than a straightforward Feedback Presentation. To start with, the diaries of the various, usually senior, users and managers need to be synchronised for a date in the near future. This quite often leads to the need to substitute some preferred attendees with deputies.

The choice of the facilitator is critical. The person should be someone not closely (if at all) connected with the project, so that there is no axe to grind. The facilitator needs a vast range of skills, including:

> Communication skills
> Negotiating skills
> An understanding of group dynamics
> Knowledge and understanding of the analysis method and techniques being used (at the level described in this book)

Some analysis and design skills
Project management skills
Knowledge of any automated tools being used during the session
A fast learner and assimilator of complex information
The ability to ask 'dumb' questions
To be completely egoless
The ability to think quickly on their feet

Clearly, no such beast exists, and so one needs to find someone with as many of these skills as possible. The key skill is likely to be summarised as 'extrovert and a quick thinker'. The Facilitator, on appointment, must spend some time understanding the project status and review the relevant deliverables. This can quite often be achieved by the Facilitator being present at one of the Feedback Presentations, especially one of the peer reviews.

The participants then need to be agreed. There should be about 7 ± 2 participants, and these must include the senior manager who is sponsoring the project, the Project Manager and the key users. The Facilitator will call on the senior manager to arbitrate on any unresolved business matters, and will call on the Project Manager (or the delegated Information Technology manager) to arbitrate on any unresolved technical matters. This final arbitration should be used sparingly as it overrides the common understanding and consensus view. In fact, it can be seen as a failure of the Facilitator to bring the 'team' together.

The Facilitator will circulate a set of pre-workshop material. This material will be as agreed with the senior manager and IT manager. The most important part of this pre-workshop material will describe the reason that the workshop is being held, the context in which it is being held and the overall objectives.

The Facilitator must also ensure that a suitable venue is chosen. It must be away from the main participants' working area, hold no telephones and be properly equipped with presentation aids, coffee facilities and side-rooms for caucus meetings. Caucus meetings should be discouraged as everything should be discussed in the open; however, managers/users sometimes like to get together in order to wash any dirty linen in private and so ensure a common objective, maybe through management control. Two key presentation aids that I have found useful are: (1) an overhead projector with plenty of blank acetates and coloured pens that work, and (2) one of those clever whiteboards which, at a press of a button, can be photocopied. The facilities must be booked for a suitable period of time; there is nothing worse than having to halt an interesting and useful discussion because the room is about to be converted into a ballroom.

The Facilitator is also responsible for ensuring that the meeting keeps going; this is sometimes harder than ensuring that the meeting keeps to a well-defined agenda. In order to do this the Facilitator needs to be armed with a set of potential discussion points and perhaps, given the skills that the Facilitator is supposed to process, a set of questions on topics which need clarification.

The Rapid Interactive Session should start with an overview of the structure and way of working for such a session. The way this is done is important as it sets the scene for the workshop. Each Facilitator will want to stamp his or her own personality onto this introduction. The following is a summary of the introduction that I have used many (many many) times. The introduction lasts for about an hour, which seems a long time in this context. However, by the end the participants have settled down, asked some jokey 'dumb' questions, know me (as the Facilitator) a little better and feel comfortable with their peers. After such an introduction they can start on the hard work that is required to complete the session successfully.

Overhead 1   Welcome to ...

Overhead 2   The agenda, my standard overhead is shown in Figure 3.1.

Figure 3.1
Rapid Interactive
Session −
introductory session
agenda

Introduction and Objectives

- Why you are here today
- How we are going to run the day
- The participants
- The topics
- My credentials
- Your credentials

Overhead 3   The theory of why the meeting is taking place, starting with the overhead shown in Figure 3.2.

Figure 3.2
The start of a set of
overheads giving the
general reason why
the interactive session
is taking place.

Why are we here today?

To agree a view of the business which everyone in the business can understand and agree to

---

Overheads 4−14 A summary of Chapter 1 (the reasons for doing modelling).

Overhead 15 This gives the direct answer to the question posed in Overhead 3. Figure 3.3 gives a real-life example I used some time ago.

THE DIRECT ANSWER TO
'Why are we here today?' is:

- To discuss key issues in the Feasibility Study, and
- To ensure that we understand the document, and
- To agree changes, and
- To ensure that there is enough information to finalise the timescale, and agree that the proposed timescale is reasonable and acceptable.

Figure 3.3
An example of the specific answer to, 'Why are we here today?'

Overhead 16 This overhead sets the scene for how the day is going to be run. My standard overhead is as shown in Figure 3.4. The asterisks on this slide indicate that subsequent slides explain them in more detail. I feel that this slide needs no further explanation here.

How are we going to run the day?

1. As a Partnership
   - everyone is equal in terms of these discussions
2. As one meeting
   - everyone should be able to have their say
   - everyone should listen to the other's point of view
3. Discuss one issue at a time
   - agree it, discuss it, record the result
4. As an open meeting
   - ask dumb questions*
   - this is the opportunity to agree the project contents
   - acquiescence will be explicit agreement
5. As a fun day*

Figure 3.4
How we are going to run the day?

Overhead 17   A quote from Tom Peters on dumb questions, shown in Figure 3.5.

Figure 3.5
Tom Peters on dumb questions.

You must have the guts to ask dumb questions

Mostly, it's the "dumb", elementary questions, followed up by a dozen even more elementary questions, that yield the pay dirt

e.g. "Why in the heck does this form go <u>there</u> next?"

Tom Peters
Thriving on Chaos

Overhead 18   Yet another quote from Tom Peters, this time on having fun, shown in Figure 3.6.

Figure 3.6
Tom Peters and having fun.

HAVE FUN

'Only fun − in the sense of taking pleasure in accomplishment and interesting foul-up alike − will allow you to thrive amidst the ravages of change in a world turned upside down!'

'The economic stakes have never been higher, therefore, it's never been more important <u>not</u> to take yourself too seriously'

Tom Peters
Thriving on Chaos

Overhead 19   A list of the participants. Figure 3.7 gives an example where some of the prescriptive techniques and names used above were slightly altered to match the organisation's way of thinking. Notice that I was the Facilitator and provided some of the analyst skills. The names of the participants have not been shown in order to protect the innocent. Note also an attempt to bring some humour into the proceedings, and thus 'break the ice' right at the beginning, by the use of 'moi' on this overhead.

Figure 3.7
A list of the participants.

The Participants	**Date** _____
FACILITATOR AND ANALYST	
• runs the day	– mol
• structures the results	
SENIOR CUSTOMER	
• the business authority	– Mr W
CUSTOMER	
• the business knowledge	– most of you
IT MANAGER	
• the technical manager	– Mr X
ANALYST/PRESENTERS	
• responsible for technical material	– Mr Y
SCRIBE ANALYST	
• documents the day	– Mr Z
• ensures understanding has taken place	

Overhead 20   A list of the topics for the workshop session. Obviously, this will change for each workshop. Figure 3.8 is a real-life example.

Figure 3.8
Real-life example of the topics of a Rapid Interactive Session.

The Topics	**Date** _____
1. REVIEW BUSINESS OBJECTIVES	
2. REVIEW SCOPE OF SYSTEM	
3. IMPLICATION OF BUSINESS PRACTICES	
4. QUESTIONS RAISED BY THE IT DPT	
5. RECAP AND FINAL AGREEMENT	

Overhead 21 An overhead introducing the Facilitator to the group, listing name, past achievements in this area and association, if any, to the project.

Overhead 22 An overhead asking for details of each of the participants. Figure 3.9 is the basic overhead that I use. The Facilitator asks each of the participants to say a few sentences on who they are, the role they play and the worries that they have.

Figure 3.9
And who are you?

> **AND WHO**
> **ARE**
> **YOU**
> **?**
>
> • Name
> • Role
> • Any worries/problems that you have with Feasibility Study

I have run a number of these Rapid Interactive Analysis sessions for a variety of organisations. Frequently, the first session is taken up by the group introducing themselves to each other and understanding their role on the project. Their description of their role quite often comes as a surprise to some of the others. Although the project may have been running for quite a while, the initial Rapid Interactive Analysis meeting may be the first time that the senior managers, senior users and IT analysts have met.

It is hard work for the analyst who *must* structure the information on-line, must ensure that the key questions are being asked and answered, and must work hard between sessions to complete the structuring of the information and ensure that information is at least consistent. The users and managers can be fairly relaxed at these sessions as they should be telling the analyst as it is, putting forward their everyday methods of working. The analyst, on the other hand, must work very hard, both during the meetings and between the meetings (quite often overnight). The Rapid Interactive Sessions will not be successful if the analyst does not work hard enough, and I have seen some sessions fail because too much old ground is covered at a subsequent meeting, mainly because not enough of the analysis work was carried out between the meetings.

One of the projects I was involved in had already been going for some while, and a number of reports had been produced. A number of key business issues had been raised and 'definitive' answers obtained. However, we all agreed that the proposed system did not quite hang together, and so it was decided that a Rapid Interactive Session would be held. I was the Facilitator at the meeting. The meeting started by everyone introducing themselves, then I presented the basic rules for the meeting. Then, in order to get the participants into the swing of things, I asked a safe, basic question about the 'agreed' scope of the proposed system. The room of about twenty people, fifteen of whom were senior users and managers, resounded to a cacophony of yes's, no's and well maybe's. It took a few, unexpected hours of the meeting to obtain a common understanding on the system we were discussing. And remember that most of the people in the room had already been interviewed and had agreed the structured information passed back to them.

3.5   Notes, Issues and Queries

This chapter has concentrated on the basic techniques for gathering and presenting information. Later chapters describe the processes that take place between the gathering and presenting: the structuring of the information.

A number of points should be remembered when analysing the information:

1. When interviewing, all of the information being passed should be noted, even if not relevant to this part of the life-cycle.
2. When presenting information, only information that is relevant to the Reviewers should be shown.
3. The information about the system should be fleshed out over time so that there is enough information for the designers to work on, and then enough for the implementers to code the system from.

As the information is being gathered, the analyst will record all of the relevant information either in a structured picture plus supporting material, or in 'Notes'. 'Notes' is a collection of all those bits and pieces which do not quite belong in the mainstream analysis, but either add additional information or are things which will be of concern later in the design and implementation phases. Such notes can range from statements on the hardware platform to acknowledging one of the problems with the current system that must be solved during the design phase.

As the analyst structures the information acquired, it will become clear that pieces of the jigsaw are missing. Structuring the information and performing some checks will lead the analyst to discover all sorts of inconsistencies, such as:

1. Everyone thinks that they are receiving a piece of data, yet no-one is supplying it (this is a very common fault).
2. No-one is taking responsibility for certain activities (usually to do with exception handling).
3. The scope of the proposed system is not well understood/defined, leading to confusion as to what current systems, if any, will be eventually replaced.

The analyst needs to consider these areas of incompleteness and inconsistencies and, within the partnership between the analyst and users, decide:

1. Whether the problem can be 'solved' by discussions within the partnership.
2. Whether the problem can be 'solved' by raising it as a technical Query for review and agreement by a specified group of project staff.
3. Whether the problem requires user agreement and/or management decision higher than the partnership can provide — in this case it is raised as a business Issue.

Hence, associated with any structured view of the business/system are a number of additional points termed Notes, Queries and Issues. The structured view itself is not just a picture. A picture in modelling is *not* worth a thousand words: it is an overview to understanding the thousand words associated with it. For each picture presented there must be at least one sheet of A4 paper explaining it.

In particular, each 'box' and 'line' on Figure 3.10 should have at least a one line description of it in the associated text. Preferably, there should be examples of the 'box' or 'line'. Also, of course, the associated Notes, Queries and Issues will be recorded.

Figure 3.10
A structured model consists of at least a picture and a page of supporting text.

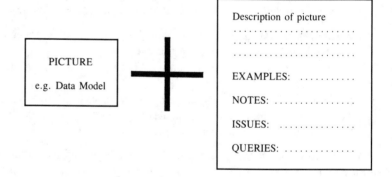

4 On Business Analysis

4.1 Introduction

Data Analysis and Activity Analysis are techniques for describing the data and functions which need to be supported by the business. Some of this support will be automated via one or more computer systems, sometimes termed Information Systems. Once an overview of the business is obtained, the analyst will normally describe support for just one part of the overall business needs, homing in on one specific area of the business, sometimes termed a Business Area.

If one asks a manager what information he or she would like in order to do the managerial tasks allocated to him/her, the likely reply will produce a long list. When asked when this information is required in the average business monthly cycle, the likely answer is 'yesterday'. I am sure that many analysts/programmers know of at least one computer system which has had many weeks of hard-toiled effort poured over it just to produce a report that no-one uses any more — and most probably never did. This is not wholly the fault of the user/manager: everyone has a long list of what one would like. There has to be some structure so that everyone in the business can decide what is needed. Business Analysis is the starting point for determining that structure — and hence what is needed.

How do we decide what the important business information needs and business activities are? There are two main issues. The first is to determine the business information needs and activities for the whole business, the second relates to the specific information needs and activities for the business system under consideration. The second is, of course, a subset of the first (Figure 4.1).

How much work is performed on the whole business is a matter of judgement. Clearly, producing a full analysis for the whole of a large organisation such as ICI or British Airways would be too daunting a task to carry out, even if there was enough time, money and staff to do it. However, the main steps of Business Analysis, as described below, should always be performed for some large area of the business which needs a set of inter-communicating support systems: for example, large areas in British Airways could be Engineering, Catering and Passenger Reservations. Once the large areas have been analysed, the results can be moved 'upwards' towards producing the analysis for the whole organisation.

The decision on what activities and data belong to a specific area of the business is a scoping issue and is dealt with in Chapter 5 on Activity Analysis.

Figure 4.1
Areas of business
within the whole
business.

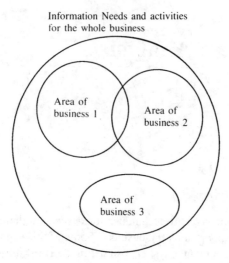

Information Needs and activities
for the whole business

Area of
business 1

Area of
business 2

Area of
business 3

4.2 Business Analysis

The gathering of Information Needs and activities for the whole business is a
company-wide issue and is sometimes defined as part of business strategy. I have
employed another commonly used term, namely Business Analysis.

In performing Business Analysis one is attempting to determine the information
needs *required* to support the objectives of the business. The information required,
not the information that might, possibly, be useful. That is, not a wish list, but
the minimum set of data which must be held in order to make the business function.
It is so easy to dive into building computer systems which may meet some understanding of the need; however, far too often the completed computer system fails
to meet the real needs of the user. The analyst, in partnership with the users and
managers, must stand back from the computer system and think about the business
that must be supported. The partnership must decide both:

1.   Are we doing the right things?
2.   Are we doing the right things correctly?

To help answer these questions, there is a set of basic techniques to aid the
analyst structure the business needs. These techniques are:

1.   State in one sentence the main 'mission' of the business — the mission state-
     ment. (Be careful here, we need a business-orientated mission not just 'to
     be highly profitable' unless, of course, that is the whole reason for the
     company.)
2.   Produce 7 ± 2 top-level business objectives.
3.   For each of the top-level business objectives, establish the next-level
     objectives, perhaps homing in on the area of business under consideration.
4.   Establish what can be measured in order to help determine whether or not
     the objective is being met.

5. Identify how each of the things that can be measured can actually be measured.
6. Identify the desirable values of the measurements: the measurements that the business is attempting to meet — the target value.
7. Consider the things which help or constrain the success of the objectives.

There is no magic here: just trying to structure the manager's view of the business, making sure that the view is reasonable and that identified objectives can have sensible targets assigned to them.

4.3   A simple example

Here is a simple example of using these techniques. The overall business is 'running the family home'. The mission statement is likely to be, 'to provide a happy home for all of the family'. Three top-level objectives are:

1. To keep down the cost of running the family home.
2. To ensure that each family member is well cared for.
3. To make the front garden more attractive than the neighbours'.

The first of my top-level objectives could be further decomposed as follows:

Top-level business objective:	To keep down the cost of running the family home
Next-level objectives:	To save electricity
	To save gas
	To shop wisely
	To vote for the right local council
	To have a mortgage with the best lender

I am sure that other examples will spring to mind immediately, but it is important to limit the number of objectives to around seven. This can be done by ensuring that each objective is at the same level. For example, the objective 'to shop wisely' is a higher-level objective than 'to minimise the use of 3 kW electric fires'. Some objectives may be better grouped together. For example, 'to save electricity' and 'to save gas' may be better grouped as one objective — 'to minimise heating and cooking costs' or 'to minimise the use of (public) utilities'. There will be no right answer, we are obtaining a list of objectives which make sense to, and are understood by, the partnership. For each objective we can then go on to find ways of measuring them. If we cannot find ways of measuring an objective, then the objective is likely to be meaningless. The objective 'to stop smoking' is easy to measure; the objective 'to think more often during the day' is very difficult to measure.

Let us consider measures for the first objective, 'to save electricity'. What can we measure that will help us decide whether or not this objective is being met? The most direct thing is 'the amount of electricity consumed'. Other things which may be of use to measure are 'the number of high-consumption devices' and 'the amount of time when no-one is in the house'.

Having determined what things can and should be measured, one now determines how to measure them. Here is a list of things that can be measured for the objective 'to save electricity':

Amount of electricity consumed
> Measured by change in meter dial reading over a period

Number of high-consumption devices
> Measured by number of 2 kW and 3 kW electric fires
> Measured by number of hours the cooker is used per week

Amount of time no-one is in the house
> Measured by the number of days holiday per year
> Measured by the number of days both adults are out of the house

The next step looks for targets for each of the identified measurements. Continuing the theme as to whether or not it is worth having any one specific objective: it is not worth having something to measure if one cannot usefully do something with the measurement. What we are looking for at this stage of analysis is things that we can measure to see if the objective is being met, and therefore something to which we can assign a target. The targets have to be carefully considered within the partnership. Sometimes a target cannot be clearly identified, and so the objective must be considered as a bit vague. Sometimes the agreed target is 'obviously' not achievable in any reasonable time period. Here are some targets for the above example:

Amount of electricity consumed
> Measured by change in meter dial reading over a period
> TARGET: Not greater than 200 units per day

Number of high-consumption devices
> Measured by number of 2 kW and 3 kW electric fires
> TARGET: Not greater than three fires in the house
> Measured by number of hours the cooker is used per week
> TARGET: Not greater than 5 hours per week

Amount of time no-one is in the house
> Measured by the number of days holiday per year
> TARGET: Greater than 10 days per year
> Measured by the number of days both adults are out of the house
> TARGET: 5 days per week, 10 hours per day

Even with this simple example, one can see some of the analysis work that needs to be done. For example, a technically good target for the measure of the number of days holiday per year in the context of saving electricity is 365. However, this is not a very achievable target. Measuring the number of 2 kW and 3 kW electric fires in the house and demanding a target of three is quite reasonable; however, is the measurement of the number of hours the cooker is in use reasonable? What is the cost of measuring such usage? Can it be measured accurately? Even if it can, is the target reasonable — surely one just uses the cooker when one needs to?

Remember that if a sensible target cannot be found, then the objective is not a sensible one as one can never tell when the objective is being met. Identifying a target ensures that the partnership has understood the objective and its measures — a target being a real-world value.

The targets, together with their means of measurement, help to determine the data that must be held in the business system. A substantial amount of data have been identified in the 'running a family home' example, some of which is:

Electricity meter reading
Number of 2 kW electric fires
Number of 3 kW electric fires
Period of time that the cooker is on
Number of days the family is away from the house on holiday
Working pattern of the adults

Associated with these objectives and targets are things that may help a particular objective succeed, and things that may constrain the success of one or more objectives. Some of these things will not be measurable, while others could be measurable but a decision is made not to measure them.

A few of the things which would help to save electricity are:

Moving to Cyprus
Always wearing woollen jumpers
Eating out a lot
Improving the insulation in the house

A few of the things which constrain the objective are:

Long, cold winters
Having one of the family ill
Having to look after your mother-in-law

Listing these 'success factors' and 'constraints' is important as it helps to establish whether an objective can be met from a more subjective viewpoint. For example, an objective 'to choose the cheapest electricity supplier' has no success factors in the UK and is, in fact, a major constraint as the supplier holds a monopoly. Hence the objective is meaningless.

4.4 Identifying problems

While the partnership is putting together the business objectives, the users and managers will tend to address a number of the problems that they are experiencing with the current business support, manual and/or automated. These problems must also be noted and then structured against each of the objectives. Then the analyst can show the users and managers how the proposed system detailed at the end of the analysis phase can help solve the problems, or why an identified problem

Figure 4.2
The business mission
statement can be
broken down into one
or more top-level
business objectives.

Figure 4.3
Each top-level
business objective can
be refined into one or
more lower-level
objectives.

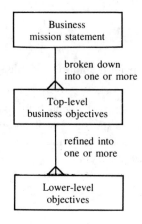

Figure 4.4 Each
lower-level objective
can be measured by
one or more
measurable things.

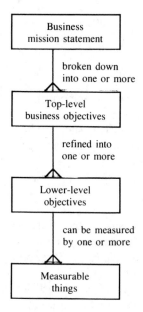

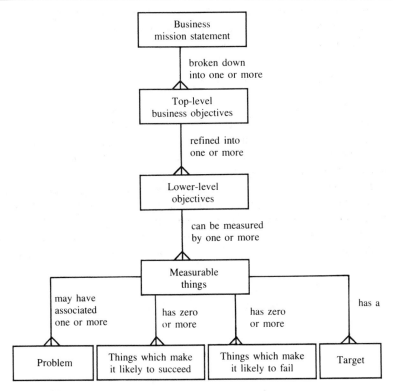

is not going to be solved — for example, because it is not cost effective to solve it.

Some of the problems that might be identified with the objective 'to save electricity', and which a new system may be able to help solve, are:

Leaving high-consumption devices on by mistake

Misreading of the meter

Not knowing how much electricity has been consumed until the quarterly bill arrives

4.5   Summary of Business Analysis

Business Analysis starts with the overall mission statement of the business. This is then broken down into a number of top-level business objectives, as shown in Figure 4.2.

Each top-level business objective is then refined into a number of lower-level objectives, moving towards the objectives of one specific area of the business instead of for the whole business (Figure 4.3).

For each lower-level objective, one decides what can be measured in order to see if the objective can be met (Figure 4.4).

For each measurable thing, the analysts, with the help of the users, find the target value for the measurement. Then the partnership determines the things which constrain and/or contribute to the success of the objective, usually directly related to each measurable thing. As the analyst structures these things, the perceived problems with the current environment are logged (Figure 4.5).

4.6   The health warning

Without knowing it, you have just read a Data Model showing the information being gathered during Business Analysis. Just as this model helps you picture what is happening in Business Analysis, so modelling your business will help you understand the information required by the business *before* trying to build computer systems to support that business.

Too often, Business Analysis is not carried out at all. This tends to mean that each computer system is built in splendid isolation, and the whole does not meet the business needs. Much effort is then expended in joining up the independent systems and providing 'fudges' in order to meet some of the needs. How right is the adage, 'Having lost sight of our objectives, we redoubled our efforts.'

5 On activity models

5.1 Introduction

There are basically three techniques used during activity analysis, namely:

1. Context Diagrams: Used to determine the scope of the area of the business under consideration.
2. Activity Hierarchy Diagrams: Used to document the business activities.
3. Data Flow Diagrams: Used to show the data flows between activities and data that have to be stored within the system.

A brief note on the use of the terms 'function', 'activity' and 'process'. Many of the prescriptive definitions of methods spend pages describing the distinctions among these three terms. The best use that I know is to use 'function' to describe things that have to be done by the top-level business, 'activity' to describe the things that have to be done by a particular area of the business, and 'process' to describe the physical task, automated or manual, that has to be carried out in order to support the activities. However, I have used the words 'activity' and 'function' interchangeably in this book, the rationale being that activity and function are descriptions on *what* has to be done. I have tended to use 'process' to mean a physical task: that is, *how* something is done physically in order to support what has to be done. Hence, the technique I have called Activity Hierarchy Diagrams could equally be called Function Hierarchy Diagrams. Some books replace the word hierarchy with decomposition, thus Function Decomposition Diagrams. Most of the techniques out there in the world are based on the basic techniques described here, even though they are all given different names. Activity Analysis is a much older set of techniques than Data Analysis, and so there are many more different techniques and variants of using them.

All three techniques, that is, the Context Diagram, the Activity Hierarchy Diagram and the Data Flow Diagram, have to be used in an iterative manner. The analyst will do a bit of the Context Diagram, map out a first-pass Activity Hierarchy Diagram and then go back and rethink the Context Diagram. Then the Data Flow Diagrams will be started, and the Context Diagram and Activity Hierarchy Diagram revisited.

Figure 5.1
Start of a Context
Diagram.

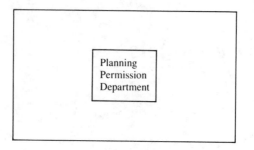

5.2   Context Diagrams

The Context Diagram is used to determine the scope of the business system. In the standard technique, a box, labelled with the business system, is placed in the centre of a blank sheet of paper. Then 'things' which relate to the business are placed around the outside. One may not know what all these things are to start with, but some things will certainly be clear. Then the data flows to and from the system are drawn in and described.

A simple example is of a Planning Permission Department within a local council. Taking a blank sheet of paper, and without waiting to be intimidated by the emptiness of the paper, draw a box in the centre of the paper and place the words 'Planning Permission Department' in the centre of the box (Figure 5.1).

What things are associated with a Planning Permission Department? The person, or persons, asking for the planning permission is a good start: let us call this person, or persons, the 'applicant'. Local planning laws will need to be considered. A mailing shot is needed to the local neighbours. Comments will have to be gathered from the neighbours and from other interested parties. These can be drawn onto the Context Diagram, as shown in Figure 5.2.

Going into more detail, the Planning Permission Department will have to interact

Figure 5.2
Part of Context
Diagram for Planning
Permission
Department.

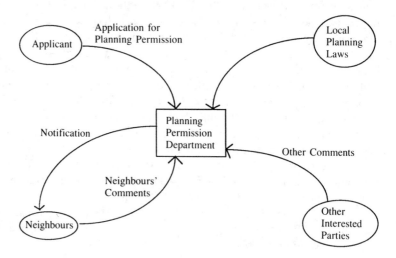

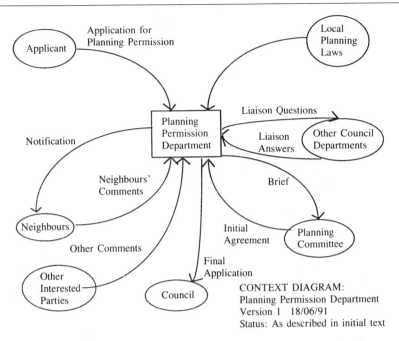

Figure 5.3
Version 1 of Context
Diagram for Planning
Permission
Department.

CONTEXT DIAGRAM:
Planning Permission Department
Version 1   18/06/91
Status: As described in initial text

with other local authority departments, such as Health and Safety, and Highways. The Planning Permission Department prepares a brief for the Planning Committee, which consists of some of the locally-elected councillors. The final decision-making takes place at a council meeting. The first version of the Context Diagram can now be put together (Figure 5.3).

As I draw this diagram, I immediately notice that two flows are missing; this was not clear when writing and reading the textual description. The diagram does not show what happens to the final application once it has been to the council — there needs to be a flow from 'Council' back to the Planning Permission Department giving the final decision. This decision then needs to be communicated to the applicant.

Reviewing Figure 5.3, I have also noticed that the flow from 'Local Planning Laws' to the Planning Permission Department has not been named. There is a school of thought that states that if a flow is not named, then that flow is everything from the described business-related thing — in this case the local planning laws. There is another school of thought that says that it is more likely that the analyst has forgotten to name the flow, and so action is needed to correct the picture. Having named the flow, the Context Diagram develops into that shown in Figure 5.4.

You can argue among yourselves whether or not the naming conventions I have used on the Context Diagram are correct. I like the use of capitalised words on Context Diagrams. It is just a matter of taste — the most important thing is to ensure that the picture helps to promote understanding about the business and the proposed systems to support the business.

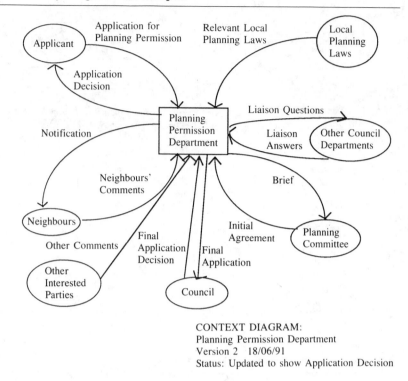

CONTEXT DIAGRAM:
Planning Permission Department
Version 2   18/06/91
Status: Updated to show Application Decision

5.3   Discovering verbs

Having considered the Context Diagram, one has a better understanding of the
activities required within the area of business under consideration. The analyst
will consider the results of the Business Analysis, further background text and
interview notes in order to discover verbs from the point of view of the (in this
case) Planning Permission Department. Each verb is a basis for some activity
in the business system.

Paraphrasing the verbs in the textual description of the Planning Permission
Department (PPD) given earlier, we have:

> The PPD *receives* application for planning permission
> (Presumably — must check) the PPD *reviews* the application
> The PPD *considers* local planning law
> The PPD *mailshots* the neighbours
> The PPD *receives* objections
> The PPD *interacts* with other departments
> The PPD *interacts* with the Planning Committee
> The PPD *interacts* with the council
> The PPD *informs* the applicant of the decision

These activities have been written in the order in which they were described in
the text, and are more or less chronological. This is usually the best way to start.

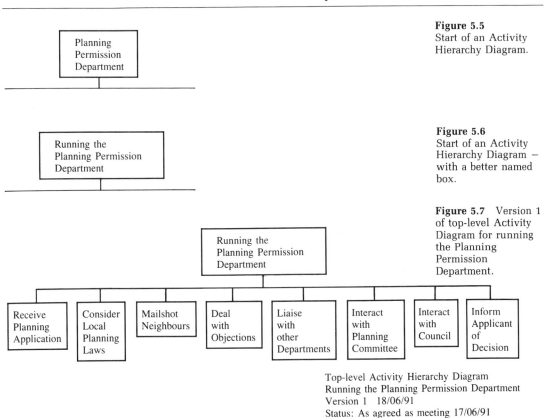

Figure 5.5
Start of an Activity
Hierarchy Diagram.

Figure 5.6
Start of an Activity
Hierarchy Diagram —
with a better named
box.

Figure 5.7 Version 1
of top-level Activity
Diagram for running
the Planning
Permission
Department.

Top-level Activity Hierarchy Diagram
Running the Planning Permission Department
Version 1   18/06/91
Status: As agreed as meeting 17/06/91

5.4   Structuring verbs into an Activity Hierarchy Diagram

These activities are now structured into an Activity Hierarchy Diagram. An Activity Hierarchy Diagram shows the activities that make up the specific area of the business under consideration. The diagram is only an overview and does not attempt to show the ordering of the activities nor the data flows. One must consider it as a brain dump of the necessary activities.

Therefore, just as with the Context Diagram, take a blank sheet of paper and, without waiting to be intimidated by the emptiness of the paper, draw a box at the top centre, placing the words 'Planning Permission Department' in the box (Figure 5.5).

However, the boxes in an Activity Hierarchy Diagram represent verbs — things that the business has to do — so it would seem more sensible to name the boxes with verbs. Hence the top box would be better named 'Running the Planning Permission Department' (Figure 5.6).

The next level of activities can now be added straight from the paraphrasing and capture of verbs detailed above (Figure 5.7).

We are not sure at this stage whether or not these are the correct higher activities. To gain a better understanding, the analyst will now decompose each activity. 'Receive Planning Application' may comprise the activities:

> Receive application in post
> Check application for completeness
> Update application if incomplete
> Acknowledge receipt

In order to obtain this decomposition, the analyst would reconsider the available background information, ask more questions and/or take some intelligent guesses. 'Take some intelligent guesses' may not sound like a proper interaction with users but, in my experience, it is very difficult to obtain all the information required at the first interview. Neither the analyst nor the user knows exactly what is required. What the analyst must do is structure the information in the best way possible and then fill in the gaps to make a coherent story. This can then be presented to the user, as a Feedback Presentation, and the user can make constructive criticisms on the work to date — having something to criticise instead of having the dreaded blank sheet of paper.

The activity 'Consider Local Planning Laws' may consist of:

> Check application for non-standard features
> Identify relevant local planning laws
> Review identified local planning laws
> Document impact of local planning laws

These two activities can now be expanded onto the Activity Hierarchy Diagram shown in Figure 5.8.

In general, the analyst will not write out the hierarchy as I have started to do in the figure, but rather draw the Activity Hierarchy Diagram correctly. Figure 5.9 shows the first-pass Activity Hierarchy Diagram for 'Running the Planning Permission Department'. This diagram is quite easy to read and understand, and may be considered to be worth a thousand words. The problem is, what thousand words are they? Each user will read into each activity whatever he or she feels is most suitable; almost certainly, this will not be the same as what the analyst believes the activity name represents. Hence it is absolutely vital that each activity has at least a one-line description associated with it, together with the notes and queries as discussed in Chapter 3, Section 3.5.

Here is a selection of the background texts for the activities shown on Figure 5.9.

> Receive application in post
>> Receive the planning application in the local council postal system. This includes applications delivered by hand, as these are handed in at the council reception desk and then put into the internal post.
> Check application for completeness
>> Check that the application has all of the mandatory boxes completed and that the relevant fee has been included.

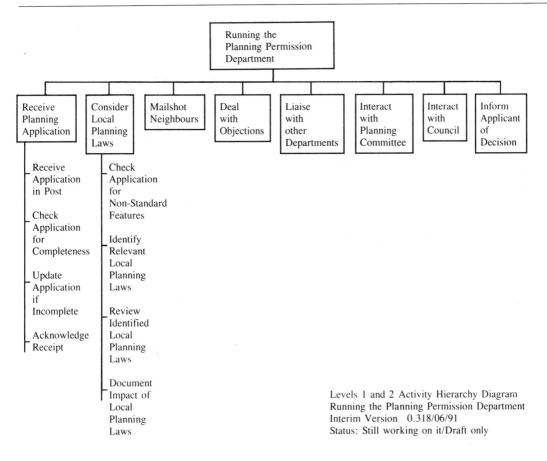

Levels 1 and 2 Activity Hierarchy Diagram
Running the Planning Permission Department
Interim Version 0.318/06/91
Status: Still working on it/Draft only

Figure 5.8
Moving towards
first-pass Activity
Hierarchy Diagram for
running the Planning
Permission
Department

BUSINESS QUERY: What happens if any of the checks fail? Are other checks made, for example, legibility and understandability?

Stuff envelopes

For each identified neighbour, obtain their full address and print out suitable envelope labels. Take the mailshot material and place each one in an envelope, add label and place through franking machine.

Receive objections

Receive objections through the internal post and check that the reference number is correct: that is, it matches a valid outstanding planning application. Check also that the objection has been received in time.

NOTES: The mechanism for checking the validity of objections is not yet clear to me.

BUSINESS QUERIES: Can anyone object? What happens to objections that are out of time?

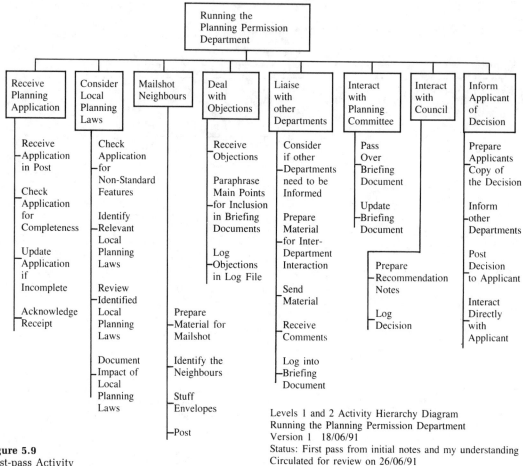

Figure 5.9
First-pass Activity
Hierarchy Diagram for
running the Planning
Permission
Department.

5.5   Data Flow Diagrams

The analyst has now produced, with the users' help, a first-pass Context Diagram and a detailed Activity Hierarchy Diagram. It is now time to obtain a more comprehensive view of the activities listed in the Activity Hierarchy Diagram. This is done using Data Flow Diagrams. Data Flow Diagrams show how data flows between the activities already identified on the Activity Hierarchy Diagram, how data flows to and from the 'things' shown on the Context Diagram, and what data needs to be stored within the system. There are three pieces of terminology that are worth learning at this stage, as most practitioners use them when detailing with Data Flow Diagrams. These are data flow, data store and external entity. (It does not matter whether these three terms are capitalised or not; all that matters is that their meaning is clear to the users, managers and analysts in your business.)

Data flow: Showing the path of pieces of data that flow through
                  the system. The flow can be between activities, from/to
                  an external entity and an activity, and from/to an
                  activity and a data store.
Data store: A holder for pieces of data that the activities using them
                  need to keep around for a period of time and/or wish
                  to share with other activities which are not necessarily
                  performed at the same time.
External entity: This is the proper name for the 'things' on the Context
                  Diagram, being the things that provide data to the
                  system and/or receive data from the system.

Note that Data Flow Diagrams can show many other things, and each specific
method has additional specific-to-method techniques. This book does not address
any of them, as most are detailed techniques for analysts and, as such, are unlikely
to be shown to users. If Data Flow Diagrams with additional detail are presented
to you as a user, then ask the analyst for a brief training session first. Make sure
that you really need to understand the level of detail being shown; if necessary,
the user should ask the analyst to simplify the picture. This chapter concentrates
on the basics.

There will be a set of Data Flow Diagrams, one for each 'arm' of the Activity
Hierarchy Diagram. The start of the first Data Flow Diagram is derived from
the Context Diagram and the top-level activities of the Activity Hierarchy Diagram
(Figure 5.10).

Now the top-level Data Flow Diagram is filled in with the relevant data flows
and data stores. The first obvious flow is 'Application for Planning Permission'
from the applicant into 'Receive Planning Permission'. What is the likely flow

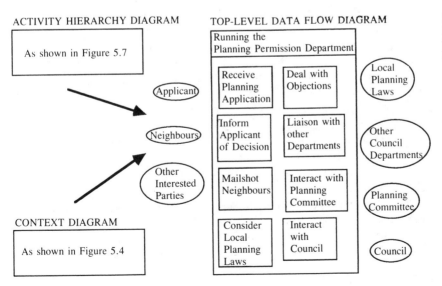

ACTIVITY HIERARCHY DIAGRAM

As shown in Figure 5.7

CONTEXT DIAGRAM

As shown in Figure 5.4

Figure 5.10
Deriving the first
top-level Data Flow
Diagram.

Figure 5.11
Starting to fill in the
top-level Data Flow
Diagram.

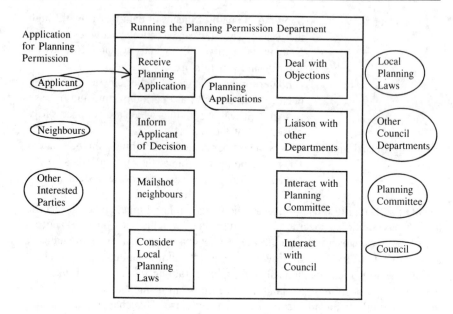

out of this activity? The fact that the department has a planning application sparks off a series of other activities: for example, informing the neighbours and then waiting for objections. Hence it would seem reasonable that the Planning Department should store the planning application plus the comments in a 'store'. Hence, let us add a data store called Planning Applications to the diagram. The start achieved so far is shown in Figure 5.11.

There is a data flow from the activity 'Receive Planning Application' into the data store 'Planning Applications'. The flow could be called 'Received Planning Application'. The activity 'Mailshot Neighbours' will use this data store to prepare and then send notification to the neighbours. Neighbours will send back 'Neighbours' Comments' and other interested parties will send back 'Other Comments'. Both these will be dealt with by the activity 'Deal with Objections'. The data store 'Planning Applications' will be updated with a summary of the objections plus a log of where the full objection can be found. The story so far is shown in Figure 5.12. (Notice the data store 'Planning Applications'.)

The analyst continues until the top-level diagram is completed (Figure 5.13). Notice that some data flows are not named. This is shorthand for 'the data flow consists of everything in the data store or external entity with which the line is associated'. I have deliberately used this convention here to show you that it is very confusing. One is not sure if the data flow is as just described or is simply not labelled due to an error in producing the diagram. I always recommend that every data flow is named.

Once the analyst is happy with the first set of activity diagrams, the diagrams must be reviewed with the users and managers. This is to ensure that the diagrams are complete and consistent. From this stable base the analyst can produce Data

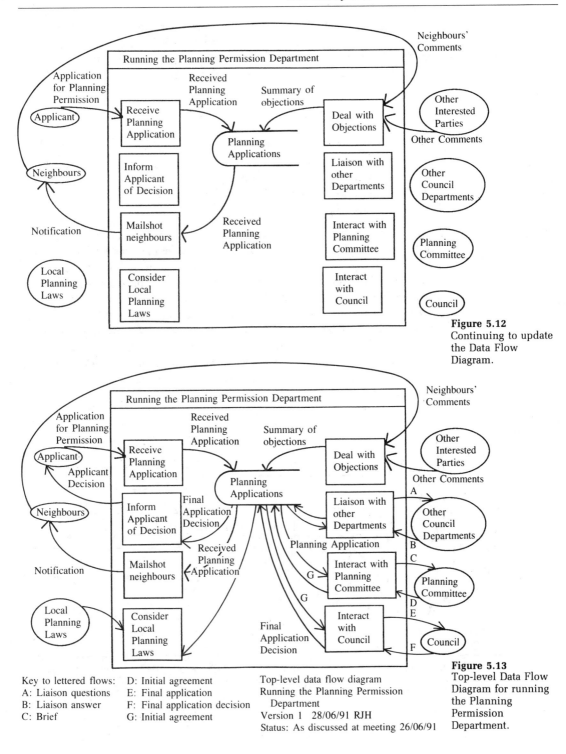

Figure 5.12
Continuing to update the Data Flow Diagram.

Figure 5.13
Top-level Data Flow Diagram for running the Planning Permission Department.

Key to lettered flows:
A: Liaison questions
B: Liaison answer
C: Brief
D: Initial agreement
E: Final application
F: Final application decision
G: Initial agreement

Top-level data flow diagram
Running the Planning Permission
  Department
Version 1   28/06/91 RJH
Status: As discussed at meeting 26/06/91

Flow Diagrams for the lower levels. There is a convention that helps to keep control of this hierarchy of Data Flow Diagrams. Each of the activities on the top-level Data Flow Diagram is numbered from 1. Hence, in our top-level diagram shown in Figure 5.13, the activities could be numbered as follows:

Receive Planning Permission	1
Inform Applicant of Decision	2
Mailshot Neighbours	3
Consider Local Planning Laws	4
Deal with Objections	5
Liaison with Other Departments	6
Interact with Planning Committee	7
Interact with Council	8

The analyst then goes into more detail down the separate aims of the Activity Hierarchy Diagram. The activities detailing 'Receive Planning Permission' will be numbered 1.1, 1.2, and so on; those for 'Interact with Council', 8.1, 8.2, and so on. The overall mapping between the Activity Hierarchy Diagram and the Data Flow Diagrams is shown in Figure 5.14.

When the analyst is presenting these activity diagrams to users and managers, the main Data Flow Diagram must be put into context by showing where it fits in the hierarchy of pictures. I find that a road map based on Figure 5.14 is most useful, together with the latest versions of the Context Diagram, top-level Data Flow Diagram and Activity Hierarchy Diagrams all being available if needed.

Figure 5.14
The overall mapping between the Activity Hierarchy Diagram and the many Data Flow Diagrams.

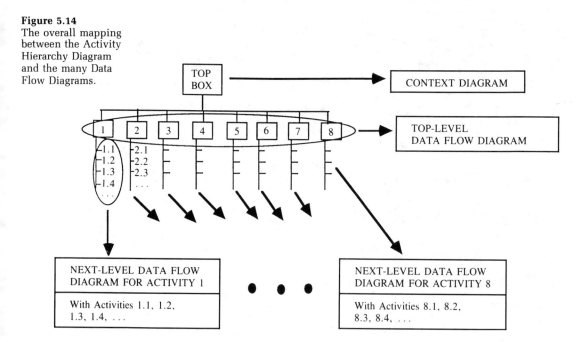

5.6   More on Context Diagrams

The standard technique for producing a Context Diagram places the business system under consideration in the middle of the diagram with the associated 'things' (external entities) around it. This is fine if only one fairly closed system is being considered and the 'things', which may include other computer systems, are well understood.

In real life this is very unusual. More and more systems have to interact with each other. Some methods, notably Information Engineering, call for a Data and Activity Model for the whole business before 'splitting' them up into business areas: the idea being that data are common across computer systems which support the business areas.

Unfortunately, this common, high-level model is hardly ever produced. Some reasons for failure seem to be:

1. The top-level Business Data Model tends not to be detailed enough at the start.
2. The top-level Business Data Model is not properly updated as further, lower-level business area models are refined.
3. Many organisations do not produce a Business Data Model, preferring to rush forward and start coding a particular computer system straight away (the WISCA syndrome — why isn't Sam coding anything? This is not the correct long-term approach. What is the point of having a lot of reasonably good computer systems if they fail to talk to each other?).

I am increasingly coming across companies that have spent a lot of money putting in computer networks, each one boasting about the latest UNIX file server, 25 MHz 386 boxes, and the latest file server software. When I ask what information is being passed over the network, the conversation dries up. Certainly, the electronic mail system has improved and the software managers spend less on expensive software as certain development tools can be shared over the network. However, it is unlikely that the end-user will be able to share data as no-one has considered how the end-user systems can share data.

Context Diagrams can be extended to help show the interface requirements among computer systems. Instead of placing one computer system in the middle of a blank sheet of paper, obtain a larger sheet of paper and place the two or three computer systems under consideration near the middle of the paper. Then add the other external entities. Then decide on the data flows. This is a very interesting exercise.

Recently, I was part of a team looking at how three computer systems would interact with each other. One of the computer systems was about to go live; the second's project team had just completed the analysis phase; and the third's project team had just completed the first-pass textual requirements specification. We placed the three computer systems on one sheet of paper together with other systems we knew we had to interact. We then did a tour of the projects to decide what data flows to add. As one should have expected, each project had different ideas about what data should be flowing between the systems and, what was far worse, in which direction it should flow. The three-system Context Diagram became a very powerful document for showing that the scope of the systems under consideration was a bit hazy and that many important data were not formally 'owned' by anyone. Not surprisingly, our report on the interfaces was fairly brief: the systems could not interact with each other.

5.7   More on Activity Hierarchy Diagrams

Activity decomposition starts from the top. One divides the business into Business Areas and then divides the Business Areas into successively smaller parts until the individual activities are of a suitable size.

The size of the 'suitable size' depends on who is to review the Activity Hierarchy Diagram. Senior managers will want to see much less detail than a system designer. However, at whatever level, the shown activity must be easy to describe in a few lines of English text.

The top line of the Activity Hierarchy Diagram is vitally important as it describes the entire set of business activities within the scope of the specification of requirements. In order for the top line to be meaningful and usable, the following points should be adhered to:

1.   There should be about 7 ± 2 business activities.
2.   These 7 ± 2 business activities should be able to be combined into one long single sentence which, by its very nature, describes the types of activity undertaken within the business.
3.   Each business activity should be named using at least a verb and a noun, for example, Order New Stock.
4.   For each identified activity, add as many lines of description as is necessary to describe the activity.

No information should be lost about an identified activity. If the information is not formally part of the description, it should be listed under the heading 'Notes'. Queries to do with the business should be listed under the heading 'Queries'.

6 On Data Models

6.1 Basic concepts

The concept of data modelling is very simple: it is to produce a structured picture of the static data that are required in order for the business to carry out the business functions. In the context of data modelling, structuring is pulling together all the pieces of information that are required in a business and placing them into a picture. The picture shows the types of information and how each type of information relates to the other types. The dictionary has one definition: 'arrangement of parts'; for our purposes, this arrangement of parts aids the understanding among the partnership of the analyst, users and managers. There are four basic constituents:

1. Entities: These are nouns that the analyst finds while reading background material and interviewing users and managers.
2. Entity Types: These are groupings of the entities in such a way that similar data are held in one group.
3. Attributes: (Sometimes called *data items*) being the actual data held within an Entity Type. This will consist of some of the entities already found plus additional data identified as now belonging to this group of data.
4. Relationships: These show how the groups of data (that is, the Entity Types) are associated with other groups of data.

The notions of Entity Types, relationships and attributes lead to the often used name for Data Models, namely ERA models. Another common name is *data structures*, sometimes qualified with a term such as 'logical'. Here is an example of determining these parts:

While conducting an interview, the analyst notes the nouns 'airport', 'car', 'truck', 'aeroplane', 'driver', 'pilot', 'passenger' and 'check-in staff'. In the context of the business being evaluated, the analyst decides that the entities that require further consideration are 'car', 'truck', 'aeroplane', 'driver', 'pilot' and 'passenger'. The other nouns are not required as they are outside the scope of the business system.

I can well remember a course I once gave to a group of senior managers in a large organisation. I was attempting to give the managers a view of what the analysts were about to do in terms of producing data and activity models. I explained about nouns and so on, and then gave them a small exercise consisting of a short description of a particular business area. They were asked to read the description and then draw a data model. A look of horror spread across their faces. I asked them what the problem was. 'We're not analysts,' they replied. I explained that all I was asking them to do was to find the nouns, group them together and then draw a simple model. 'But that's too easy ...!' 'Remember,' I said, 'the first pass doesn't have to be correct.' Whatever correct means; the model is there to provoke discussion, making sure that the analyst understands the business, and that the users and managers feel that the analyst has understood the business.

With some thought, the analyst groups these entities into two groupings as follows:

> Entity Type = 'Transport'
>> being a grouping of 'car', 'truck' and 'aeroplane'
> Entity Type = 'Person'
>> being a grouping of 'driver', 'pilot' and 'passenger'

The analyst now considers the entities grouped within each Entity Type in order to find the first-pass attributes of the Entity Types. In the case of 'Transport', one needs to identify whether it is a 'car', 'truck' or 'aeroplane', and so a reasonable attribute is 'type of transport'. Similarly, for 'Person' there would seem to be a need for an attribute to determine what sort of person it is: 'driver', 'pilot' or 'passenger'. A reasonable attribute is 'role of person'.

Then the analyst decides that there is some form of association between 'Person' and 'Transport'. Such relationships could be 'drives', 'rides in', or perhaps 'is booked as role of'. The actual relationship will, again, depend on the business area. Let us assume that this is right at the start of the analysis and we just want to record that there is some sort of relationship. In this case, the easiest thing to do is to call the relationship 'is associated with'. This relationship will be refined later in the analysis.

In summary, then:

> A 'Person' *is associated with* a 'Transport'

There are two other important concepts, namely:

1. **Identifier:**   This is a set of attributes from one Entity Type whose combined values makes a set of data unique. (There is a complication whereby relationships may also be included in the identifier set, but I shall not dwell on this here.)
2. **Instance:**   This is a set of data, values of the attributes, describing one particular Entity Type, uniquely identified by the identifier. This is sometimes called an *occurrence*.

The following examples should make these terms clearer:

The *identifier* for 'Transport' has to be a set of attributes whose values will identify one particular *instance* of 'Transport': that is, uniquely identify a real-world piece of transport. Looking at the different 'types of transport' (being an attribute of 'Transport'), one can determine the following:

car: Uniquely identified by registration number in England.
        (BUSINESS QUERY: Do we need to consider non-English registrations? If so, what additional information is required to make a car unique?)
truck: Same as 'car'.
aeroplane: Uniquely identified by registration number for most countries.
        (BUSINESS QUERY: Is this okay?)

'Registration number' seems, then, to be a good identifier. However, is registration number sufficiently unique across the different types of transport? For safety, it may be better to add the attribute 'type of transport' to the identifier. Hence, the identifier for 'Transport' is: 'type of transport' + 'registration number'.

'Registration number' was not listed as an attribute for 'Transport' and so now it must be. Hence, considering how to identify an Entity Type uniquely has helped the analyst to decide some of the business data that need to be held.

The analyst would now have to do a similar exercise on 'Person'. This is much harder. In the UK, there are a number of ways of identifying a person, depending on the area of 'business'. For example, here are a number of possible identifiers associated with some 'business areas':

Prison system:	Criminal record identification number
Driving:	Driving licence number
Health service:	National Health Service number
Workplace:	Employee identification number
Bank:	Account number + surname + initials + selected parts of the address
Credit agency:	Address
Telephone company:	Customer account number or telephone number

Any person may have one or more of these identifiers.

Let us take the easy path for now and state that in this business area every person is an employee of the company and so has a unique employee identification number. Therefore, the identifier for 'Person' is 'employee id'.

In attempting to identify the identifier for 'Person', the analyst has found a range of attributes such as 'name of person' and 'telephone number', which can go alongside the already determined 'role of person' and 'employee id'.

The analyst can now confidently draw an initial Data Model as shown in Figure 6.1, which shows that a person is associated with a transport. More complicated relationships are dealt with later, but this will do for now. Quite often this is the first view of the data that will be shown to senior users.

Figure 6.1

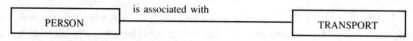

Each Person is associated a Transport

Now let us consider an instance of these Entity Types. In order to define an instance we need to give the identifier of the Entity Type its 'real world' values plus state the likely values of the other defined attributes.

For 'Transport', an instance could be the company car with the following attributes:

  Identifier attributes: Type of transport = 'car'
          Registration number = 'G768 DPL'
  Other attributes: No other attributes identified yet

For 'Person', an instance could be as follows:

  Identifier attributes: Employee id = '1'
  Other attributes: Name of person = 'Roger Hipperson'
          Telephone number = '0276-683244'
          Role of person = 'Passenger'

Some of you may be questioning the grouping of the data. For example, should 'Person' be so all-embracing? Do we need an Entity Type for 'pilot/driver' and one for 'passenger'. This would allow a more general use of 'Person', for example, to hold just the person's name and telephone number. It is this sort of thinking which structuring the data sparks off, and ensures, via the analyst's skill and experience, that the required business data is identified and its use documented. There are no absolutely correct answers, just an intuitive feeling that the information shown meets the overall business needs as understood at the time. It is the experience of the senior analysts, that is, someone with some fifteen to twenty years' experience, that helps the more junior analyst to shape the structured model in a way that helps spread the common understanding. Remember the chess story in Chapter 3.

In summary, the minimum that the user/manager should expect the analyst to provide is as follows:

1. A picture showing Entity Types and named relationships.
2. A list of attributes that make an instance of an Entity Type unique (the identifier).
3. A list of other interesting attributes.
4. Some real-world examples of instances, giving the values of the identifier plus those of other interesting attributes.

The users should be encouraged to produce some instances of their own in order to prove that the groupings are 'correct' and that there is a common understanding of the groupings.

6.2   More on relationships

The first Data Models drawn by the analyst will most likely show simply that
there is a relationship between two specific Entity Types. As the analysis continues,
the relationships can be further refined in order to show:

Cardinality
Optionality

I use these words only because everyone else does and so one should be aware
of them. I will now explain what they mean.

6.2.1   *Cardinality*

Cardinality shows how many instances of one Entity Type are associated with
instances of another Entity Type. As the analysis continues, more work must be
performed to consider how the associations between Entity Types behave.

For example, in Formula One motor car racing each instance of a racing car
is associated with one instance of racing driver. This is known as a one-to-one
relationship. There are many different notations for showing this, and one would
have to check the Project Manual for the one used on any particular project. A
common notation for this one-to-one relationship is shown in Figure 6.2.

Figure 6.2

Each Racing Car is associated with one and only one Racing Driver
Each Racing Driver is associated with one and only one Racing Car

On further thought, each racing driver is likely to be associated with more than
one racing car; the driver will only be driving one car at any one time, but there
are likely to be spare cars set up specially for the driver.

What the analyst has to do (with the help of the users) when deciding on the
type of relationship, known as cardinality, is to consider how the association looks
from either end. That is, consider the association first of all from the point of
view of one of the Entity Types, then from the point of view of the other. Hence,
for the racing car:

Each 'Racing Car' is associated with one 'Racing Driver'
Each 'Racing Driver' is associated with one or more 'Racing Cars'

A common notation for this one-to-many relationship is shown in Figure 6.3.

Figure 6.3

Each Racing Car is associated with one Racing Driver
Each Racing Driver is associated with one or more Racings Cars

The other type of cardinality is many-to-many. Before the 1991 recession, an independent consultant would have many client companies, and companies would use many different independent consultants. This many-to-many relationship is shown, using a common notation, in Figure 6.4.

Figure 6.4

Each Independent Consultant is used by one or more Companies
Each Company uses one or more Independent Consultants

6.2.2   Optionality: may or must

In the above examples, each relationship is such that there is at least one instance of one Entity Type associated with at least one instance of the other Entity Type. In real life this is not always true. A classic example is as follows:

A 'Woman' may have one or more 'Children'
A 'Child' must have, as its mother, one 'Woman'

Not all women have children, but all children have a mother, where mother is defined here as the woman who gave birth to them. A common way of representing this is shown in Figure 6.5.

Figure 6.5

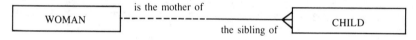

Each Woman may be the mother of one or more Children
Each Child must be the sibling of one Woman

6.2.3   Resolving many-to-many relationships

Here is an example to help explain how relationships depend on the viewpoint and how many-to-many relationships can hide/lose information that needs to be stored within the system. The example is estate agents and the selling of houses; Figure 6.6 shows a Data Model from the estate agent's point of view.

Figure 6.6

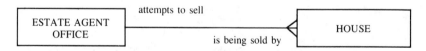

Each Estate Agent Office attempts to sell one or more Houses
Each House is being sold by one Estate Agent Office

The relationships are 'must', as we can assume that if an estate agent office is attempting to sell no houses, it will be closed down. Furthermore, from the estate agent's point of view, a house is only known about if it is being sold by the agent.

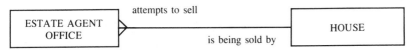

Figure 6.7

Each House is being sold by one or more Estate Agent Offices
Each Estate Agent Offices attempts to sell the one House

The picture is completely different from the house owner's point of view and this view is shown in Figure 6.7.

Let us now look from the point of view of a global house-selling system. Estate agent offices attempt to sell many houses and each house is being sold by many agents. This is a true many-to-many relationship, as shown in Figure 6.8.

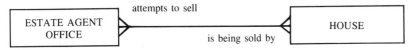

Figure 6.8

Each Estate Agent Office attempts to sell one or more Houses
Each House is being sold by one or more Estate Agents

Now, what information is one likely to need in this transaction? Here are some obvious data:

       Vacant possession?
       Selling price
       Agent's commission

Where can these attributes be held? 'Vacant possession?' is a one-off attribute of 'House' and so can be stored there. 'Agent's commission' is likely to be agent-specific and so can be stored with the Entity Type 'Estate Agent Office'. The attribute 'selling price' could be held with 'House', provided that each agent is trying to sell the property for the same price. This is unlikely, and so we need to hold a piece of information associated with 'House' *and* 'Estate Agent Office'. This could be called 'advertised price'.

Consider Figure 6.9. The house '5 Cardwell Terrace' is associated with three

Figure 6.9
An example of some
advertised prices

estate agent offices, each attempting to sell the house for a different price. The estate agent office 'Carser, Farnham' is associated with more than one house, namely '5 Cardwell Terrace' and '15 Back Street'. Figure 6.10 shows how this could be modelled.

This method of thinking is required to complete the analysis and so pass on

Figure 6.10

Each Estate Agent Office attempts to sell houses at one or more Advertised Prices
Each Advertised Price is the attempted selling price by one Estate Agent Office
Each Advertised Price is the price for one House
Each House has one or more Advertised Prices

enough details of the specification to the designers. However, note that the picture and description with Figure 6.10 are not as easy to understand as those associated with Figure 6.8. Hence the analyst must be careful about what is shown to the users and managers. In some cases, both pictures are needed, with some careful 'hand-holding' required to explain the resolution of the many-to-many relationship.

The general rule for the analyst is that by the end of the Analysis Phase all many-to-many relationships must be resolved into the pair of one-to-many and many-to-one relationships (Figure 6.11). This can make the final Data Model very difficult to read, but does ensure complete understanding.

Figure 6.11
Resolving many-to-many relationships.

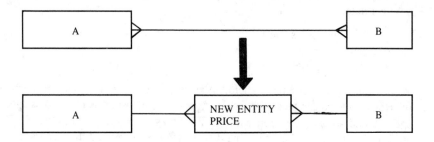

6.2.4 *Exclusive relationships*

In the UK, a student may be at a primary school, secondary school, sixth form college, a polytechnic or equivalent, or at university. From the government's point of view, each of these types of institution needs tracking in its own right, and so are Entity Types. The start of the Data Model is shown in Figure 6.12.

At any one time the student must be at a primary school, *or* at a secondary school, *or* at a sixth form college, *or* at a 'poly or equivalent', *or* at a university. Each institution will have one or more students.

In order to show the 'or' on these different relationships (called exclusivity), the most common notation is to draw an arc through the relationships — with its centre on the common Entity Type. Hence, the model started in Figure 6.12 becomes the one shown in Figure 6.13.

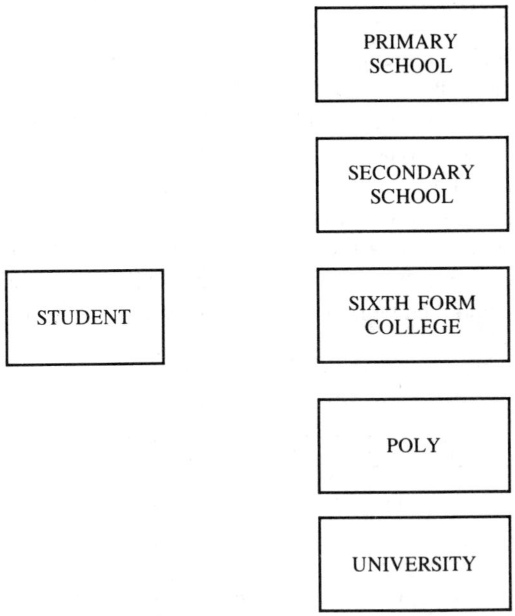

Figure 6.12 Start of the Data Model for 'Student' and 'Place of Learning'.

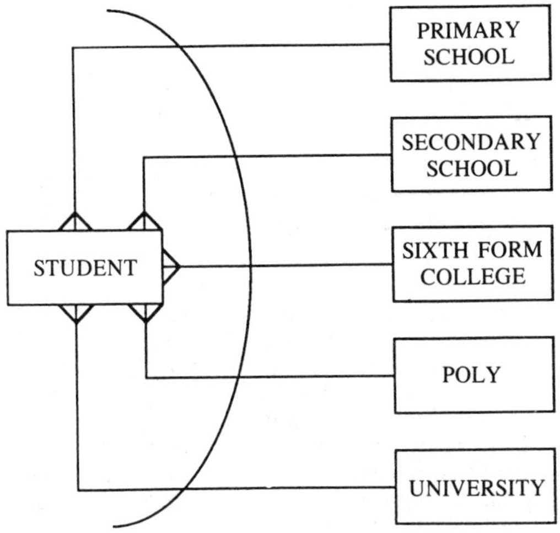

Figure 6.13

Each Student is associated with one Primary School
or one Secondary School
or one Sixth Form College
or one Poly
or one University
Each Primary School is associated with one or more Students
Each Secondary School is associated with one or more Students
Each Sixth Form College is associated with one or more Students
Each Poly is associated with one or more Students
Each University is associated with one or more Students

6.3   On subtypes

Analysts quite often simplify a picture by introducing the concept of subtypes. In the previous example, Figure 6.13, it might be decided that most of the information associated with the possible places of learning is of the same sort, for example, number of places, type of entrance qualification, location and types of final qualification. Other information, such as types of grant, funding locations and reporting structures may be different.

The analyst could decide to draw the five places of learning together into a super-type 'Place of Learning', with the already identified places of learning as subtypes. Figure 6.13 could then be redrawn, perhaps with further information,

Figure 6.14

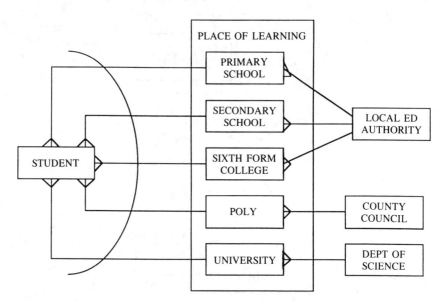

Each Student is associated with one Primary School
or one Secondary School
or one Sixth Form School
or one Poly
or one University
Each Primary School is associated with one or more Students
Each Secondary School is associated with one or more Students
Each Sixth Form College is associated with one or more Students
Each Poly is associated with one or more Students
Each University is associated with one or more Students
The Entity Type 'Place of Learning' has subtypes Primary School
Secondary School
Sixth Form College
Poly
University
Each Primary School is funded by one Local Education Authority
Each Secondary School is funded by one Local Education Authority
Each Sixth Form College is funded by one Local Education Authority
Each Poly is funded by one County Council
Each University is funded by the Department of Science

as shown in Figure 6.14. (If some of you exclaim, 'but the local education authority does not deal directly with schools any more because of local management of schools, so the picture is wrong', all well and good. It shows that the picture has helped to express my understanding of the situation and has allowed you (1) to understand by understanding, and (2) constructively to criticise it.)

Now, 'Student' is linked to all of the boxes (subtypes) in the Entity Type 'Place of Learning', and so the picture could be simplified, if it helps understanding, to the model shown in Figure 6.15. The analyst and user would have to decide if Figure 6.15 conveys the same strength of feeling as Figure 6.14 does, and whether that strength of feeling is really necessary.

Figure 6.15

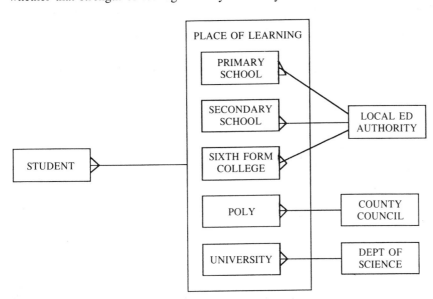

Each Student attends one Place of Learning
Each Place of Learning is attended by one or more Students
The Entity Type 'Place of Learning' has subtypes Primary School
                                                 Secondary School
                                                 Sixth Form College
                                                 Poly
                                                 University
Each Primary School is funded by one Local Education Authority
Each Secondary School is funded by one Local Education Authority
Each Sixth Form College is funded by one Local Education Authority
Each Poly is funded by one County Council
Each University is funded by one (the) Department of Science

6.4 On drawing Data Models that are easy to read

Although it may seem clever for the analyst to draw one model with between one hundred and two hundred Entity Types on it, such a model will be nigh impossible to understand by anyone else (including, I suspect, the analyst after a few months). Of course, some business areas will require this sort of number of Entity Types to be associated with it, but the key to good modelling is to split the model into many models, each with no more than twenty to thirty Entity Types on it.

My general rule of thumb is to always consider how to present the model on a standard A4 overhead projector to a group of about seven reviewers. This means that the largest sheet size would be A3 (two pieces of A4 side-by-side). A3-sized models are suitable for reports, provided that you have enough person-power to fold and insert the models into the final document! Whatever the size of the model, there are a number of simple layout rules which, when followed, greatly enhance the readability, and so understandability, of the model. These rules are:

1. Try to move Entity Types around the paper such that the number of relationships that cross are minimised.
2. One-to-many relationships are shown down the page, as shown in Figure 6.16.

Figure 6.16
One-to-many down
the page.

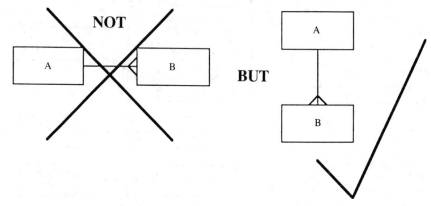

3. One-to-one and many-to-many relationships should be drawn across the page, as shown in Figure 6.17. Resolving a many-to-many will, then, be redrawn as shown in Figure 6.18.
4. If possible, the most important Entity Type for the picture should be placed in the centre of the picture and drawn larger than the others.

Figure 6.17
One-to-one and many-
to-many across the
page.

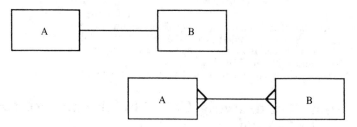

As an example of some of these techniques, Figure 6.19 shows a Data Model as it might have been constructed. Figure 6.20 shows the model tidied up somewhat, with the main Entity Type being 'Person'.

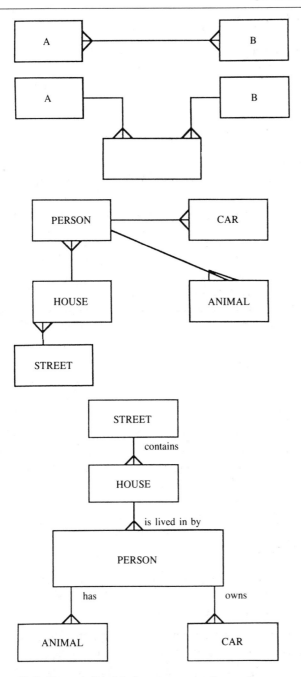

Figure 6.18
Drawing a resolved
many-to-many.

Figure 6.19
Not a very well laid
out Data Model.

Figure 6.20
The tidied-up Data
Model for 'Person'.

(Each House is lived in by one or more Persons)
Each Person lives in one House
Each Person has one or more Animals
Each Person has one or more Cars
Each Street contains one or more Houses

6.5   Manipulating the data

The analyst will attempt to put together the initial Data Model as fast as possible, usually within a few days of gathering the background information. Naturally, this initial model will need further refinement before it can be agreed within the partnership. The users and managers need not be too concerned with the actual techniques used by the analyst to refine the Data Model. They will need, though, to be aware of the sorts of thing that will change on the Data Model over the lifetime of the project.

There are three common changes made to a Data Model. The first is the resolution of many-to-many relationships into a pair of one-to-many relationships; this has already been addressed above. The second is the use of subtypes to simplify the picture and so aid our understanding of the picture; this has also been described already. The third change is to do with Entity Types and attributes. Should an existing Entity Type with a number of attributes just be a set of attributes within another Entity Type? Should a set of attributes within an existing Entity Type be split off into a new Entity Type?

This third area of change occurs as the analyst, and the partnership, gain a better understanding of the data that need to be held. For example, at the start of the analysis it was decided that the Entity Type 'Purchase Order' would have the following attributes:

> Purchase order number
> Data purchase order raised
> Name of person raising purchase order
> Name of retailer

Further analysis decided that the retailer's address should also be recorded, and so a new attribute 'retailer's address' was added. Then it was decided that retailers have a range of information that needs to be held, information that is relevant just to the retailer. The analysis has already identified 'address'; other data that became known were:

> Discount rate
> Contact name
> Telephone number

Hence the analyst can see that the details of the retailer should really be stored in a new Entity Type in its own right. The new Entity Type, 'Retailer', will have a relationship with 'Purchase Order' as follows:

> Each 'Retailer' must be the recipient of one or more 'Purchase Orders'
> Each 'Purchase Order' must be fulfilled by one 'Retailer'

The analyst would need to ask about the business issues, 'Will the system need to hold data on retailers who have not (yet) fulfilled a purchase order, and how long must we keep information on purchase orders?' Consideration of these issues may lead to the 'musts' in the descriptions above being replaced by a 'may'. If the system must always keep information on a retailer, whether or not a purchase order has been placed with the retailer, then there may not be an association between one instance of retailer and any instances of purchase order:

Each 'Retailer' *may* be the recipient of one or more 'Purchase Orders'

Clearly, a purchase order must still have one retailer. Or does it? Here is another business issue that requires clarification: 'Does one purchase order only go to one retailer?' It would seem sensible that it does, but it is worth double-checking.

On a similar vein, the analyst may have decided that, in a garage system, information held on customers would include their name, address, car registration number and type of car. During subsequent discussions it is found that quite a few customers have two cars, and so the attribute list is updated to include car registration number for the second car and type of car for the second car.

The key business issue is then to decide whether two cars per customer is sufficient over the lifetime of the system: in fact, is there a cast-iron maximum?

It is easy to miss asking these questions, and the analyst must spend time chipping away at them. I do not know what went wrong with the British Telecom billing system in early 1991. When the government changed the rate of VAT, the billing system was unable to put out the correct bills for the period in which the VAT changed. Naturally, it put out bills at the higher VAT rate. I suspect that the analyst's model only allowed for one VAT rate during a billing period when, in fact, the upper limit of changes is not known, being a whim of the government.

If the upper limit of the so-called repeating group of attributes is known (but get a senior manager's signature), then it is most probably easier to have them within one Entity Type, thus keeping the Data Model less cluttered. If the upper limit is unknown, the analyst will move the repeating group of attributes into a new Entity Type. And so in the simple garage model the Entity Type 'Customer' is split into 'Customer' and 'Car', as shown in Figure 6.21.

This process of expanding one Entity Type into two can happen in reverse, where two (or more) Entity Types are collapsed into one. Take, for example, a patient logging system for a mobile chest X-ray unit. The analyst, coming from a strong health service background, decides to model the information as shown in Figure 6.22.

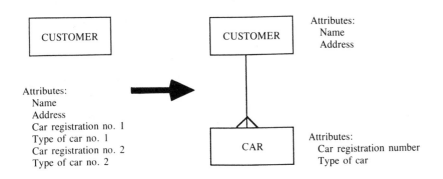

Figure 6.21
Moving repeating groups into new Entity Type.

Figure 6.22
The initial model for
the mobile chest
X-ray unit.

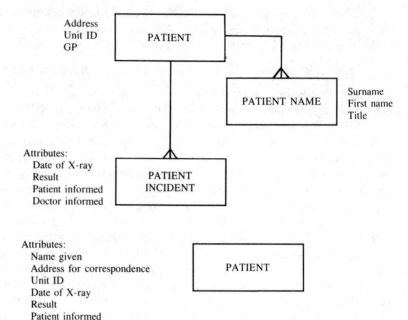

Address
Unit ID
GP

PATIENT

PATIENT NAME

Surname
First name
Title

Attributes:
  Date of X-ray
  Result
  Patient informed
  Doctor informed

PATIENT
INCIDENT

Figure 6.23
The final model for
the mobile chest
X-ray unit.

Attributes:
  Name given
  Address for correspondence
  Unit ID
  Date of X-ray
  Result
  Patient informed

PATIENT

The model allows patients to change name, for example, by getting married, and allows for more than one X-ray to occur and be logged against the patient. When this model is reviewed, it is clear that the mobile X-ray service is a one-incident service. If the results of the X-ray are 'positive', then the unit gets in touch with the patient who must then contact his or her own doctor. Hence the model can be greatly simplified, as shown in Figure 6.23.

This last example is also a good advertisement for KISS (keep it simple, stupid). The model in Figure 6.22 is more 'elegant' than the final one shown in Figure 6.23, but Figure 6.22 is *wrong* for this particular area of the business.

7 On reconciling the Activity and Data Models

7.1 Introduction

Once the set of structured deliverables based on the techniques described in earlier chapters has been produced, the analyst must check and double-check to see if the different views of the system are consistent and, as far as can be determined, complete. The basic checks, and so the minimum that should be done, are:

1. Ensure that the data held in the data flows and data stores in the Data Flow Diagrams can be obtained from the Data Model — determined by a tick matrix, as explained later.
2. Ensure that the data held in the Data Model are, in some way, used by the activities in the Activity Hierarchy Model — determined by a CRUD matrix, as explained later.
3. Ensure that the data held in the Data Model are, in some way, owned by the organisational units as shown in an organisation hierarchy — determined by a tick matrix.
4. For the key Entity Types, discover what happens to the data items within the Entity Type from its creation to its deletion — determined by the technique of Entity Life Histories, as explained later.

In general, the analyst will not present these reconciliation views to the users and managers, although they may be used to aid the analyst to obtain missing information during an interview.

7.2 Using matrices

Matrices are used to compare and contrast two parts of the total model. There are two basic styles:

1. The tick matrix, whereby a tick is placed at the intersection of two items which correspond in some way.
2. The CRUD matrix, used to determined what activities create, read, update or delete a particular piece of data.

An example of a tick matrix may be demonstrated when matching data in a data store with attributes stored in the Data Model. For example, the Data Model may have three Entity Types with the attributes listed below:

Person	Name
	Address
Car	Make
	Version
	Engine size
Worksheet	Date
	Service details
	Standard price
	Quoted price
	Extras
	Actual price
	Invoiced price

The data store in the Data Flow Diagram called 'Work-in-Progress' contains the following data items:

> Date-of-work
> Price structure
> Car details
> Owner details

The analyst needs to check if the information required in the data store can be obtained from the Data Model, and so draws up a tick matrix as shown in Figure 7.1.

The analyst finds that the data store can obtain the information from the Data Model, but wonders why the data store 'Work-in-Progress' does not store anything to do with the Data Model attribute 'service details'. A technical query will be raised and the matter discussed.

The other matrix technique, the CRUD matrix, is extremely important as it allows the analyst to do the following:

1. To identify which pieces of data are not created and/or not updated and/or not read and/or not deleted. After all, it is no use one bit of the system trying to read an attribute which no part of the system is creating.

2. To identify which groupings of data and activities should be implemented together, the idea being to group activities and data in such a way that the main 'users' of data are implemented together. Several methods have techniques for mathematically grouping the activities based on the relevant weightings of the create, read, update and delete values in the matrix. This book does not attempt to describe how the analyst performs this operation: it is enough at this stage to understand that the analyst will be determining such things and that users and managers are likely to be questioned over a period of time in order for the exercise to be completed.

DATA STORE: 'WORK-IN-PROGRESS'				
ENTITY	Date-of-work	Price structure	Car details	Owner details
PERSON:				
name				✓
address				✓
CAR				
make			✓	
version			✓	
engine size			✓	
WORKSHEET:				
date	✓			
service details				
standard price		✓		
quoted price		✓		
extras		✓		
actual price		✓		
invoiced price				

Figure 7.1
Tick matrix for data store 'work-in-progress'.

In the garage example above, the following activities have been determined:

> Log-in car
> Perform work
> Log-out car

An example CRUD matrix is shown in Figure 7.2. In this simple example, there are neither updates nor deletes. It would seem reasonable not to be updating anything, but the analyst would need to know when the data held in these Entity Types is going to be deleted. Perhaps a new activity is required, such as 'flush out old records', to be run once a week. The analyst can raise a business issue asking how long the data need to be kept.

Figure 7.2 shows the low-level matrix with each of the attributes shown. Earlier in the analysis, the analyst is likely to prepare this checking matrix comparing activities against the Entity Types only. It is this higher-level matrix that can also help to group the data and activities together.

Figure 7.2
CRUD matrix for top-level garage activities.

ENTITY / ACTIVITIES	Log-in car	Perform work	Log-out car
PERSON:			
name	C		R
address	C		R
CAR			
make	C	R	
version	C	R	
engine size	C	R	
WORKSHEET:			
date	C	R	R
service details	C	R	
standard price	C		R
quoted price	C		R
extras		C	R
actual price			C
invoiced price			C

7.3   Entity Life Histories

The concept of Entity Life Histories is to take a specific Entity Type and work out what happens to instances of the Entity Type from their creation to their deletion.

Analysing what happens to a piece of data is extremely important as it: (1) helps one understand what happens to a piece of data, and (2) it shows how a piece of data moves from one system to another. This latter point helps to define the interfaces between the systems making up the computerised support, and between the automated support and the manual procedures.

It may be that certain things only happen very infrequently and so they are not worth automating. This is no excuse for not documenting such things. Once documented, the project team can then ensure that such things are considered by the manual systems supporting the automated systems.

The Data Model shows the static data that must be kept. The data relate to a number of different levels of understanding:

1. Data directly related to the identified 'Information Needs', that is, data required to support the business from a manager's point of view.

2. Data which must be held in order that the data described at (1) can be supported.
3. Data which must be held to support the lowest-level activities discovered during the Analysis Phase.
4. Additional data required to support design features such as integrity, on-line testing, historical records, etc.

Entity Life Histories take a process view of the data and so serve to clarify:

1. When a new instance of an Entity Type is created.
2. What events/processes update the stored data (attributes).
3. What values any piece of data is expected to have at any one time/period.
4. Circumstances in which a data item can be deleted.

It is worth noting that we are obtaining a view of the data contents of an Entity Type instance over time. We are not changing the structure: that is, the number and type of attributes is constant.

There are a number of techniques for drawing Entity Life Histories, the most common being Jackson Structures. Jackson Structures are used to define all of the things that can happen to an Entity Type without referring directly to activities identified on the Activity Hierarchy Diagram. Hence it is a free-ranging analysis technique utilising the terminology of the users who manipulate the data. The technique is well described in the SSADM 4 manual listed in the Bibliography, and in many other publications. Its strength is that, in drawing a Jackson Structure, all of the options are raised, even if the analyst does not know what occurs in each of the cases. Hence, drawing Jackson Structures usually leads to quite a lot of discussion.

Figure 7.3 is an extract from a real Entity Life History, put together to show what happens when a cheque is sent to someone for payment and it is either cashed successfully or, for some reason, to be listed, is not cashed successfully. The

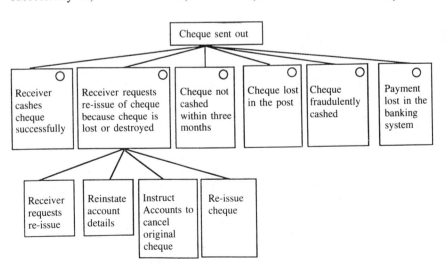

Figure 7.3
Extract from an Entity Life History.

diagram should be read as follows. Here is a part of the Entity Life History to explain what happens when a cheque is sent out as a payment. One of six things can happen:

1. The receiver cashes the cheque successfully.
2. The receiver requests re-issue of the cheque because, although he received it, he subsequently lost or destroyed it.
3. The cheque is not cashed within three months of its being sent out.
4. On investigation, it is shown that the cheque has not been received because it was lost in the post.
5. The cheque is fraudulently cashed.
6. The payment is lost within the banking system.

Only the second case has been further decomposed. If the receiver requests the re-issue of the cheque because, although he received it, he subsequently lost or destroyed it, then the following actions/events take place in the order shown:

1. The receiver requests the re-issue of the cheque.
2. The account details of the receiver are reinstated.
3. Accounts are instructed to cancel the original cheque.
4. Accounts re-issue the cheque.

8 On project hygiene

8.1 Introduction

This book is primarily about performing practical analysis. However, analysis cannot take place in an environment where no consideration has been made of the overall project structure. There are many aspects of this and I will touch on a few of them. In particular, this chapter reviews:

1. Project Planning.
2. Version Control of the models.
3. Standards and the need for the Project/Quality Manual.

8.2 Project Planning

This book does not go into too much detail on Project Planning, although, of course, these are important aspects of performing analysis. Nevertheless, at the start of any project it is important to spend some time deciding who should be seen, how long the different phases of the project are likely to take, who should sign off each phase, and what the important milestones are. All these should be agreed at the meeting that formally initiates the project: the Project Initiation Meeting. At this meeting the initial schedule, signed and published by the senior manager sponsoring the project, should be circulated. In this way, the managers and users will:

1. Know when they are going to be interviewed.
2. Know when they are likely to participate in a review.
3. Know when all of their inputs have to be ready by.
4. Know that a senior manager agrees to the workplan and so agrees to their time being spent on the project.

The top-level milestones are based on 'user acceptance' and 'installation'. The project will be given approximate timings for these, such as 'we require this system within the next three years'. Hence, a top-level project plan can be drawn fairly quickly. Figure 8.1 shows the skeleton plan awaiting the dates to be pencilled in.

Figure 8.1
Overall project plan.

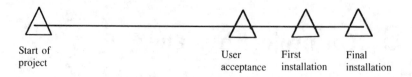

Start of   User   First   Final
project   acceptance   installation   installation

8.2.1   Project phases

A typical plan will have a number of major milestones based around the ends of each of the project phases. The simple life-cycle used in this book would have the following major milestones:

1. Project Initiation Meeting.
2. Sign-off Business Analysis Phase and initiate the Activity and Data Analysis Phase for one area of the business.
3. Sign-off Activity and Data Analysis Phase and initiate the Design Phase for one or more systems within the area of business analysed so far.
4. Sign-off Design Phase and initiate the Implementation Phase.
5. Sign-off the Implementation Phase and decide on the next steps.

This set of milestones would normally be drawn as shown in Figure 8.2.

Figure 8.2
Project plan showing
project phases.

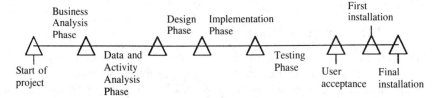

The overall plan is agreed with the senior manager, most probably during discussions on the likely project: that is, well before the initial formal Project Initiation Meeting.

At the start of each project phase a much more detailed plan would be produced for that phase. An overview showing the milestones within the phase would be produced for the users and managers. Typically, the key events to be placed in the diary are:

1. The start-date of the project phase.
2. The dates of the key interviews — in order that both parties can think about the issues and be prepared for a concentrated session of fact gathering.
3. The first date for senior managers and/or users to review the initial findings, usually in a one-to-one situation.
4. The first Feedback Presentation when the results are formally presented to a small group of managers and users for agreement.
5. The sign-off meeting where the senior manager formally agrees the work performed during this part of the project and decides on the next part of the project.

A much more detailed plan would be produced for the project team, detailing at least the following:

1. The tasks that need to be performed.
2. The effort required for each of the tasks.
3. The resources allocated to each of the tasks.
4. The elapsed time for each of the tasks.
5. The deliverables expected from each of the tasks.
6. The quality required.

The exact nature of these detailed plans will depend on the project control method being used and, to a larger extent, the automated support being used. The details of controlling a project at this level is outside the scope of this book.

At the start of each project phase in the case study presented in Part 3 of this book, I have produced the user/manager project plan for that phase.

8.2.2   Project Initiation Meeting

The most important part of the whole project is the Project Initiation Meeting. This meeting sets the scene for the whole project and, in many ways, sets the scene for the success or otherwise of the project. The meeting should consist of the proposed project team, the senior managers and the senior users. There need not be any limitation on the size of the meeting.

I remember well the first such meeting I went to in the USA. I was a Team Leader on a large military system project. Everyone was at the meeting: the whole software project team from England, the whole hardware team from America, the Department of Defense Project Management Team, and senior users from all branches of the military who may or may not use the system about to be developed. The meeting lasted most of a week. The important thing was that, although there were over a hundred people there, everyone had a chance to mingle and hear each other's views. I was in charge of two parts of the system, the low-level operating system and the general-user interface support software (this was in the days before the phrase graphical user interface was coined, and certainly before the advent of Gem and Windows on PCs, and X-Windows on UNIX). The key message I took away from that meeting was hammered home one evening when I was having supper with a senior US Marines officer. 'It is all very well what you are proposing for the screen inputs, Roger,' he drawled, 'but remember the average user of this system has an IQ of two above plant life.' It put all my clever design work into perspective.

There are three main aims of the Project Initiation Meeting:

1. For managers, users and the project team to meet each other.
2. For the managers and users formally, and informally, to present their views of the project. This allows the users to feel, quite rightly, that they have a say in the project, and so have a feeling of ownership.
3. The most senior manager has an opportunity to address the whole project — users, managers, analysts, designers and implementers. Hence, the manager

can state the support the senior management team is going to give to the project and urge everyone to participate fully.

One of the best analysis projects I worked on was where the Project Initiation Meeting consisted of a whole day of presentations by the users. The users were invited to present their ideas for the new system, openly admitting their problems and the like. I had only been involved on the project for one day and so listened to all of this with a fresh mind, and managed to meet all of the managers and users that I had to interact with over the coming months. The best part, though, was that the managing director sat in on most of the meeting and gave a speech giving his full backing to the project and encouraged everyone to participate. This was one of the talking points of the meeting, as the managing director did not appear that often, and for some of the staff this was the first time they had heard him talk. This senior support certainly made my life easier when I started telephoning around for interviews, points of clarification and setting up Feedback Presentations.

8.2.3   The trust between the users and the analyst

The user must understand the project timescale so that information can be passed to the analyst in a timely manner. The analyst must work in order to meet the stated timescale so that interviews and reviews are performed on the date laid down and kept free by the users and managers.

A project which lays down a sensible set of interaction points and then meets them with the correct deliverables is a healthy project. The users will have faith in the analysts and continue to help them. A project which is forever changing dates and/or produces less than useful deliverables will soon degenerate into the old-fashioned scenario where users do not trust the analysts, the analysts end up 'designing' the system on their own, and the users end up with a system that they do not want and certainly cannot use.

The partnership between the user and the analyst is essential, and both parties have to work hard in order for the partnership to work. The benefit is that the system is 'owned' and 'wanted' by the users.

8.2.4   Getting the level of detail right

One of the hardest things to get right is deciding how much detail to put into the product for delivery. There are always a number of conflicting goals:

1. The need to document the understanding of the requirements.
2. The need to make the deliverables understandable to the current target audience, and yet detailed enough to move forward to the next target audience.
3. The need to meet, in some way, the existing project schedule.
4. The need to meet the analyst's perceived view of the necessary standard for the style of the material (the 'give me another day and I'll improve it' syndrome).

A general expression to remember when resolving the conflict is:

> Cut your coat according to your cloth.

That is, recognise what the most important things are and do them in the best possible way with the available resources and within the available time.

I quite often hear the cry 'but we can't perform that part of the method as it'll take too long'. On investigating further, I find that the analyst is trying to follow a prescriptive technique to the letter instead of using some pragmatic intelligence. The analyst needs to decide what parts of the specification require full analysis to the level detailed by the method. In this book, Part 2 describes the basic techniques that should be widely used; Part 3 describes more techniques which should only be used in order to flesh out certain areas of the specification. These certain areas include:

1. *Management hot buttons*. Management hot buttons are areas which the users and managers have identified as being the most important aspects of the system, and therefore those aspects which decide whether the system is politically a success or not. Management hot buttons are easy to identify as they will be mentioned again and again at interviews and presentations.

2. *Areas least understood by management*. Managers and users will not fully understand all of the areas of the proposed system. However, some areas such as interfacing requirements to other systems, especially those outside the manager's area of business, and technology support considerations, do require a high level of understanding from everyone in the partnership.

3. *Areas least understood by the analyst*. Quite often, structuring the information gathered will raise more questions than it clarifies. Most questions can be resolved fairly quickly, but there will be other areas that require more analysis in order:
   - to discover the questions that need to be asked;
   - to ensure that the analysis is complete and well understood.
   These areas are the hardest to find because the analyst, usually without malice, hides the lack of understanding behind his or her ego and 'obvious under-standing of the processes involved'. It is leaving these areas in an unclear state that causes problems during the design phase. Clearly, if the analyst cannot fathom the requirement, the designer and implementer will be unable to produce a system to support it.

4. *Areas of concern*. Even if the area is well understood by both the users and the analyst, there will be occasions where not enough is known about the area to ensure that it can be automated in a suitable manner. Hence, more work must be performed, quite often moving through parts of the design phase for this area.
   One important implementation aspect which requires this additional work is the database implementation. Designers need advance warning on database sizing and rate-of-use of parts of the database. Managers also need the information as it affects the cost of the system.

5. *Areas of change*. One of the biggest mistakes made by method designers and project teams is to assume that, once a project starts, then the requirements for the project will stay static. No matter what the size of the project, outside influences will almost certainly change some part of the specification before the first version of the software is released. This appears to fly in the face of the need to plan the project and structure the requirements. However, the set of plans and structured requirements is the objective baseline from which the team can decide what the scope of the required changes are, and so accurately cost the implementation of the change both in terms of money and on the timescale.

The analyst can alleviate some of this work by attempting to identify the potential areas of change as the project progresses. Two common areas are:

- proposed changes in government legislation;
- review of potential technology changes — in 1991, for example, this was in the areas of user interfaces (Windows 3 on PCs) and multi-user networking on PCs.

A healthy project will perform at least one task looking at the impact of technology over the expected life-time of the system being produced.

The most important thing to remember is not to allow the project to be paralysed by analysis. It is rarely necessary to dot all the i's and cross all the t's for all of the techniques described in the prescriptive method/set of techniques being followed. Unfortunately, there is no hard-and-fast rule for getting the level of detail right. The skilled analyst must use common sense and experience in order to decide what is right for each particular project. Remember the chess story in Chapter 3.

8.3   Version control of the models

8.3.1   *Latest is not always the best*

I spent a distressing time with some UNIX and C programmers some years ago. No sooner had a plan and set of program modules been agreed (reluctantly, I must admit, on the programmers' side) than the programmers would go and change everything — the new way being 'much better' than the previous thoughts. Programming on the fly may be fine for small projects (less than two people, and even then I doubt it). It has been shown time and again that it does not work for large projects (more than one person).

Any project that needs to be controlled in terms of: (1) meeting a client's wishes (even if — especially if — those wishes are changing); (2) a budget, and (3) a timescale, requires a sound baseline from which to move to the next baseline. Each baseline is as important as the next/last. Management and users need to know what has happened during the move between baselines.

This is just as important during the Analysis Phase. Earlier baselines are usually more understandable by users and managers than later ones. This is especially true of Data Models. Each diagram/narrative which has been formally reviewed should be kept in the project library for future reference.

This way of thinking applies equally as well to upgrading reviewed deliverables for correction, for improved understanding and/or a new requirement. Again, management wants an audit trail to help explain why things have changed, what the changes are, and why the deliverables now cost more and are being delivered later. This 'historical log' also helps protect the analyst from unjust criticisms.

Finally, we may go down the wrong path. Some new analysis work may show that the implementation will cost more and take an unacceptable (to the users) time to implement. In this case, the project needs to backtrack to a baseline that was acceptable and then move down an alternative path.

8.3.2 Fully label deliverables

During any project there will be a large number of deliverables, some fully documented in bound volumes, others less formally distributed for presentations and review. It always amazes me how few of these documents have any form of labelling on them; and yet analysts always expect their reviewers to have the correct documents in front of them. In fact, I have been at quite a few Feedback Presentations where the reviewers have one set of documentation, which was up-to-date 'last Wednesday', and the presenter presents the latest version — 'being the latest and so the best'; and confusion reigns.

It is *essential* that a project defines a set of rules for labelling deliverables. The minimum set of identification material that I have found essential is:

1. The date agreed for release of the deliverable. (*Not* the date the deliverable was printed out — which seems to be what so many ill-designed automated support tools produce.)
2. Some form of version identification, e.g. 1.3.
3. The status of the deliverable, e.g. produced for review on date.
4. An indication of the project phase, e.g. Design Phase.

It may be useful in a larger project to identify the team that produced the information. Again I have been amazed, especially in large organisations, how 'shy' analysts are when considering putting their name to a deliverable. The point is that the analyst *does not own* the deliverable but is its guardian, while everyone reviews it and takes *joint ownership*.

On a more detailed level, it is extremely important to ensure that the narrative for the diagram and any associated documents are well labelled, making it clear that the set of documents belong together. There are a variety of ways of doing this and the Project Manager should make a way of doing it clear at the start of the project. A technique that I sometimes use is summarised in Figure 8.3, which is a simple scheme with the association shown in only one direction. It would be better to have the descriptions referenced from the Data Model as well; however, this becomes a nightmare to keep up-to-date and would require some form of model administration.

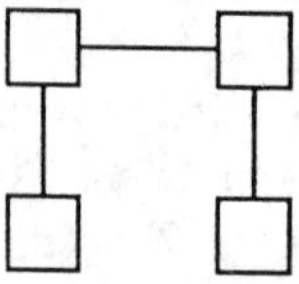

Figure 8.3
A simple technique
for labelling
deliverables.

Data Model for System B
Version 2.3   Date 3/9/90
Status: As reviewed by Jones 2/9/90

Entity Type Descriptions for System B
Version 4.9   Date 3/9/70
For associated Data Model, Version 2.3
Project Phase: Activity and Data Analysis

. . . .

. . . .

. . . .

. . . .

8.3.3   The need for model administration

Information is an extremely important commodity in any organisation. Users manipulate data in order to perform their tasks, managers use data to help make their managerial decisions. Many of the data held within one part of an organisation must be shared with other parts, in order that the different parts of the organisation can work together, and in order to provide consistent data to management.

Data and Activity Models will be produced within a number of projects within any organisation, and each will almost certainly overlap with others. Such overlapping will not only indicate shared data/activities, but also that one system may feed data into another, and/or may require data from another.

Using existing models *before* a project starts gives the analyst a sound basis to start the analysis work. The analyst can then spend more time understanding the business issues, resolving the technical queries and promoting a common understanding of the business need. By presenting high-level models of the proposed system early on in the project, the managers and users can feel that they really are a part of the partnership, and all parties can identify and resolve the areas of uncertainty and not waste time 're-inventing the wheel'.

Hence, intellectual effort is put into building a system which matches the managers' and users' needs, interfaces correctly with other systems, and produces a system that offers no surprises to users.

Model administration helps support the sharing of models in a number of ways, particularly:

1.   Model and Model Version Control.
2.   Identifying existing models which may be of use to the project.
3.   Ensuring a sound data interchange between different systems.

Model and Model Version Control

As already stated, there are many different models around in any one project. Quite often the analyst deems the latest model to be the most important and the only version available. In fact, there is likely to be at least three important versions of a model, namely:

1.   The current officially signed-off version.

2. The previous officially signed-off version.
3. The version currently being worked on.

The differences between these three versions can be very important: for example, they may show how the thinking within the organisation has changed during the discussion on the business requirements.

There are likely to be variants of the 'official' models; subsets of the models used to present certain aspects of the whole system to particular audiences. It will be important: (1) to remember what was shown at the last meeting (or whatever), and (2) when showing a new version, to explain what changes have been made.

In summary, there are a lot of versions and variants of models around. Each needs careful identification and safe storage throughout the project. The latest version of a model is not always the most important and is hardly ever the only version that is required.

Identifying existing models

Before setting out on any analysis exercise, it is always worth checking whether some of the work has not already been done. Quite often, though, the analyst will be put off searching for other models; some of the reasons are likely to be:

1. The manager does not allow the analyst time to visit other parts of the organisation in order to review other models.
2. The analyst is a bit wary of letting on that a key part of the system is not going to be modelled in-house (ego?).
3. The task of trawling through the many models may be far too daunting.
4. The 'Not Invented Here' syndrome is strong in most of us, and our culture for sharing data is very poor.

Ensuring sound data exchange between different systems

Unfortunately, most analysis methods are prescriptive techniques for a single project/system. The days of the stand-alone system are numbered and most new systems have to interact with other systems, existing or otherwise. In order for systems to interact, one must, at least, ensure that:

1. The data required and the data generated by the system are well defined.
2. The transformation processes for the data are defined adequately to ensure that there are not unnecessary, repeated activities between systems.
3. The data are presented to the systems requiring the data in a timely manner.

Any one project will find it very difficult to ensure that these data-interface requirements are consistent and complete; it needs an overview model of the interacting parts of the systems. This overview model will highlight the shared data and any necessary 'interface' functions. One important aspect of the shared data will be its ownership: who is responsible for each item of shared data and how this responsibility changes over time (if at all).

Summary of the roles of a Model Administrator

The Model Administrator must ensure that the key needs for data and activities as outlined above are carried out within the projects and between projects. In order to do this, the Model Administrator must hold a central repository for all Data and Activity Models being produced, be party to understanding them (for example, in a quality assurance role) and ensure that the projects, in some way, are taking responsibility for the interface requirements.

In order to do all of this successfully, each Model Administrator will need to have access to the models available to other Model Administrators. There is most probably a need for a hierarchy of Model Administrators, starting with one per project 'reporting to' one for the organisation.

This control of the many versions of the model deliverables is, of course, a configuration control/management problem. Techniques for doing this in a well-controlled manner, let alone correctly, are beyond the scope of this book.

Most of all, though, the Model Administrator must champion the need and the implementation of a data sharing and data exploitation culture within the organisation. This can only be done by leading by example and so helping individual projects to move forward, taking into account the important aspects of sharing information.

8.3.4   *Some notes on layering models*

Chapter 5 described Data Flow Diagrams (DFDs) and how they are layered from the Context Diagram (level 0) through a number of layers each containing the magic 7 ± 2 activities. Activity Hierarchy Diagrams can be similarly viewed with lower-layer detail for any particular function hidden on higher-level views of the diagram. Unfortunately, no-one, to my knowledge, has come up with a usable way of 'layering' Data Models. Certainly, some toolsets allow for subsets of the Data Model to be used/displayed — but these subsets are literally just some of the Entity Types and relationships taken from the diagram.

As previously stated, the high-level Data Model is a view which can be understood by managers, users and designers. In fact, it is the analyst's prime role to ensure that this is so. The analyst is desperately trying to limit the number of Entity Types and only shows the main relationships. It is essential to make the model easy to understand and as uncluttered as possible, although this does not negate the need to document fully the model in the associated narrative.

This means that some 'short cuts'/'liberties'/'little white lies' are used to simplify the diagram, for example:

1. Many-to-manys which are predominantly one-to-many are shown as one-to-many.
2. Many-to-manys which show clearly what is needed (even though it is likely that these will be 'resolved' into two one-to-manys later).
3. Only major relationships are shown, with the naming from the one-to-many direction.

4. Subtyping is used only if it makes the diagram clearer (i.e. unlikely to want more than two or three subtypes, and relationships will almost certainly go to them).
5. The diagram is grouped into major areas so that presentations can be made in a structured manner.

Although the diagram itself is simplified, the supporting narrative is *not*. All information that we have gathered on each Entity Type and each relationship must be defined in the Entity Type write-ups (and relationship write-ups if such information cannot appear sensibly within an Entity Type narrative).

Once the high-level Data Model is agreed, it should be put under strict version control and:

1. Used as the baseline for continuing modelling.
2. Used as the management view of the model.
3. Used as a 'context' diagram when reviewing the lower-level detail.

Hence, as the Data Model is fully detailed, the configured high-level Data Model must be reviewed and kept up-to-date; however, the detailed Data Model should have precedence over the high-level model.

The high-level model is aimed specifically at common understanding at all levels; the primary role of the detailed model arrived at during analysis is to ensure that all of the data have been modelled (as Entity Types, attributes and relationships) and that all volumetric data have been addressed. This means that the model will very closely match the tables required in a relational database.

8.4   Standards and the need for the Project/Quality Manual

This book has presented the basic analysis techniques using a notation fairly common in the UK. However, there are many other notations and many more techniques.

The Project Manager, with the help of the senior analysts and designers, will have to determine the set of techniques that will be used on the particular project. Due cognisance will have to be taken of at least the following, given in no particular order:

1. Standards already in use within the organisation.
2. Emerging national and international standards that the organisation ought to be moving towards.
3. The client's wishes.
4. The skill level of the managers and users involved in the project.
5. The skill level of the analysts, designers and implementers involved in the project.
6. The standards used in the analysis, design and coding of other systems which could be 'creatively swiped' to help with this project.
7. The automated support available.

The Project Manager must also consider Project Control, Version Control and the format of the deliverables.

This book is not going into any detail on how all of this is done. Much help can be obtained from public standards and user groups (in the UK at least), in particular:

1. The CCTA and the NCC publish books and papers on the use of CCTA-sponsored techniques such as SSADM (an analysis and design method) and PRINCE (a project control method).
2. The BSI produce the BS 5750 (Part 1) standard for quality and the DTI produce documents such as 'TickIT' to help the software industry implement BS 5750.
3. User groups such as the SSADM User Group, the PRINCE User Group and the BCS Software Quality Management Specialist Group all provide useful forums for discussion.

The key deliverables of the Project Manager's first activities must be: (1) an overall project plan, and (2) a Project Manual. The Project Manual, sometimes called a Quality Manual, lays down the standards and procedures that will be used on the project. In particular, it must state what parts of any prescriptive methods are to be used and what notations will be used.

Managers and users should use the Project Manual as the key document for beginning to understand the project deliverables; the analysts and designers should use it in order to ensure a consistent set of deliverables, a set that the managers and users will feel that it is worth spending the time to understand.

Part 3

Analysis: case study

9 Introducing the case study

9.1 Introduction

The case study in this part of the book is based around a small computer training company called RJH Computer Training. This fictitious example owes its existence to work performed by several people for British Airways, together with direct experience of putting together training courses. The example used here is much refined from that original work, making it specific to the purposes of this book. The original work has also been extended to allow the case study to move from analysis to the later phases of Design and Implementation. This is to give users and managers a flavour of how the results of the analysis are used eventually to build computer systems. The associated programming scripts have been included to help remind analysts of the transitions between the phases.

9.2 Notes on RJH Computer Training

9.2.1 Background

RJH Computer Training (hereafter known as RCT) is in the business of providing training courses, both public and customer-specific, in various aspects of computer technology. As RCT has no training accommodation of its own, courses are held either on clients' premises or in hotels. In either case, it is necessary to ensure that a suitable room is available and properly equipped; for certain courses, some computer equipment may also have to be provided, either by RCT or by the client.

The course structure is highly modular; while public courses follow a set pattern, in-house courses are tailored to match clients' requirements and may bring together modules originally produced for different courses. It is quite possible that a course tailored for one client may subsequently be given to another client, and may even be offered as a public course.

Since the course structure is modular, care must be taken to ensure that the correct tutor and student material is provided each time a course is given, and that it is in the right place at the right time. For each module, a variety of materials will be required, possibly including computer software. Some clients prefer to do their own copying; they must be sent an up-to-date master in good time for each course.

The major areas of the business are course administration, course pricing and scheduling, public course registration, course administration, training consultancy and other consultancy. RCT's management requires comprehensive management information, particularly about costs and income, but also about the balance of work in each area and the responsiveness to customer requests.

9.2.2   Course preparation

Course material is prepared according to a modular structure. Each module is designed to last from one hour to one day, and normally includes presentation material and exercises. In general, around half the module duration will be taken up by exercises, though this does vary depending on the nature of the subject. Presentation material normally takes the form of overhead projector slides, which are produced directly on one of RCT's laser printers using a PC-based desktop publishing system. Master copies of handouts are produced in the same way; they include copies of all diagrams used on the slides, and expanded explanations of the points made on any text slides. For each exercise, both a question and a sample solution must be produced.

It is also necessary to provide explanatory notes for the person giving the course. These must cover both administrative details and technical points which are likely to cause difficulties. For some courses, computer software may also be required; this may be simply copies of existing packages or new data files for use with existing software, or sometimes new programs developed specifically for the module.

9.2.3   Course pricing and scheduling

The main variables in setting the price for a course are the duration, venue and equipment required. In addition, some courses are priced more highly to recover unusually high development costs. These calculations give a basic public course price; the price of on-site courses is based on this, in such a way that an on-site course will usually be cheaper for the customer if more than six students attend. For on-site courses outside of the London area, travelling and subsistence are also charged at standard rates; the price quoted to the customer in such cases is a single inclusive figure.

Where one or more modules are developed specifically to meet a particular customer requirement, an additional charge is normally made to cover the development costs, either on a fixed cost or time-and-materials basis. RCT's standard agreement gives both parties rights to use the resulting training materials, and modules developed in this way are often incorporated into future public courses or courses for other customers; in such cases, provision may be made for the payment of a partial refund to the original customer.

A customer may sometimes ask for the rights to use RCT's training materials for in-house courses not run by RCT. Each such request is handled by individual negotiation; the resulting agreement usually covers courses given at a single site, but on occasions may be effective nationally or even worldwide.

The public course schedule is reviewed and published twice a year and covers courses for a twelve-month period. When a course is fully booked well in advance, an additional course will often be scheduled.

On-site courses are scheduled at clients' request; the period of notice varies widely, but is usually between one and four months. In some cases, a regular programme is agreed and a discount may then be given. Where long-distance travel is involved, every effort is made to schedule courses to run consecutively to reduce travelling costs and inconvenience.

The major factor in scheduling is the availability of instructors. Premises are not usually a problem, since RCT has long-term agreements with a number of hotels and conference centres and can usually find suitable accommodation. For courses where RCT has to provide computer equipment, its availability may also be a constraint, since the company uses special networking hardware which is not readily available on short-term rental.

Most courses require instructors with substantial practical experience, and no instructor is allowed to give a course without first taking it as a student; there are no exceptions to this rule. For the first few presentations of a course, then, there may be only one or two instructors qualified to give it, usually those who did the preparation; for popular courses, this number will then increase quite rapidly. RCT's Structured Analysis course, for example, can now be given by all instructors except possibly recent recruits, and they will attend it within three months of joining.

It is RCT's policy that no instructor should spend more than half his or her time giving courses. The balance may be spent in course preparation, in training consultancy and planning, and in other specialist forms of consultancy.

9.2.4   Public course registration

While some courses are run on-site for particular customers only, much of RCT's business originates in attendance of one or two individuals from a customer at a public course. Even though this directly accounts for less than a quarter of RCT's total revenue, it is important because it often leads to further business of other kinds. Public courses are therefore regarded as a marketing activity as well as products in their own right.

Course frequencies vary between monthly and once or twice per year. It is important to review advance bookings and attendance frequently, and to adjust the schedule when necessary. Those courses which are run frequently are subject to cancellation if there are less than four bookings two weeks before the start-date; for less frequent courses, though, no course is cancelled until the requirements of each customer already booked have been considered.

9.2.5   Course administration

For public courses, RCT is responsible for all administration. This begins with the booking of accommodation and any equipment required, not forgetting meals and refreshments. Nearer the time, the course materials must be assembled and copies made for the students. Care must be taken to use the latest version of all course materials, as changes are constantly being made.

The materials must be transported to the course venue the day before the course is due to start, and arrangements checked. Any equipment required must also be checked. The administration task continues as students arrive: each must be welcomed, registered and given a copy of the course folder. During the course, any problems must be dealt with. Lastly, course review forms must be collected and analysed, any equipment cleared away and checked, and any problems with the course material recorded for later correction.

For public courses, this administration work is normally handled by one of RCT's course administrators, who will be present certainly at the beginning and end of the course, and often from time-to-time between. These administrators also handle the course bookings (see above), and are trained to advise on the suitability of courses and on alternatives to the recommended prerequisites.

For on-site courses, much of the administration is normally handled by the customer. When necessary, though, RCT will provide all the necessary computer and other equipment for a course, and it is then treated in the same way as a public course. Since the only member of RCT staff present at an on-site course is often the instructor, it is even more important that all of the materials are assembled and checked in advance. Where the customer is to produce the students' copies of the handouts and exercises, an updated master must be sent well in advance.

9.2.6   Training consultancy

RCT provides a training advisory service for customers. This ranges from free advice on the most suitable existing courses to meet particular requirements, through the adapting of existing courses with some new material, to the development of completely new courses. The development of new or tailored courses is often combined with other consultancy work (see below), particularly on methods and standards.

Where a customer has an existing training department, arrangements can be made to train suitably-qualified staff in that department to take over the presentation of the courses, while RCT will continue to provide updated material from time to time, and to provide back-up in case of illness or other emergency.

9.2.7   Other consultancy

In order to provide a comprehensive range of training, RCT employs a number of highly-qualified consultants, primarily as instructors. Recognising the importance of keeping technical knowledge up to date, RCT encourages its consultants to accept a limited number of consultancy assignments in their specialised fields. This applies in several areas, but particularly in the development and use of systems analysis and development methods and standards, an area in which RCT has built up a considerable reputation.

9.2.8 Management information

In a business such as this, it would be easy for costs to get out of control. RCT's management therefore requires comprehensive reporting against budgets on a monthly basis, and forecasts for the year are revised on a quarterly basis. Control is particularly important in course preparation, where keeping to project plans is vital to avoid delaying scheduled courses; here, reporting of costs must go down to the module level.

For each course given, whether public or on-site, the direct costs and resulting revenue must be reported. Wherever possible, costs are allocated directly to the relevant course-related activity or consultancy assignment, but inevitably some costs have to be treated as overheads. While overhead costs are normally managed by monitoring against budget, management also requires the ability to obtain reports in which all overheads have been apportioned between courses, course offerings and consultancy assignments on an agreed basis. This basis of overheads allocation may be different for different categories of cost: for example, most office costs are allocated in proportion to direct staff costs, but the cost of computer equipment is allocated more directly to those using it.

9.3 Overall project plan

At a number of preliminary meetings, the following overall timescale for the initial project was agreed. This timescale will be presented at the Project Initiation Meeting, the date of which is shown on the timescale.

Project Initiation Meeting	MILESTONE	1 Feb. 90
Business Analysis Phase	5 weeks	1 Feb.–7 Mar.
Sign-off Business Analysis Phase	MILESTONE	7 Mar. 90
Activity and Data Analysis Phase for one area of the business, expected to be in the area of scheduling public courses	3 months	8 Mar.–7 June
Sign-off Analysis Phase	MILESTONE	7 June 90
Design Phase for one trial implementation	1 month	8 June–9 July
Sign-off this part of the Design Phase	MILESTONE	9 July 90
Implementation Phase for Design area	2 weeks	10 July–23 July
Sign-off this part of the Implementation Phase	MILESTONE	23 July 90

Evaluate implemented system	4 weeks	24 July–17 Aug.
Project Meeting to decide next steps	MILESTONE	31 July 90

This plan would normally be shown diagrammatically (Figure 9.1).

Figure 9.1
Overall project plan
for the case study.

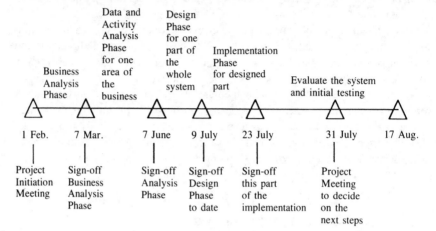

Overall Project Plan
23 January 1990, RJH
Status: For review at PIM on 1 February

10 Case study: Business Analysis Phase

10.1 Detailed plan for the Business Analysis Phase

The detailed plan for the Business Analysis Phase is shown in Figure 10.1.

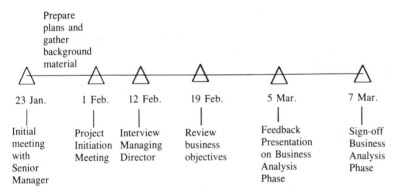

Business Analysis Project Plan
23 January 1990, RJH
Status: For review at PIM on 1 February

Figure 10.1
Detailed plan for the Business Analysis Phase.

10.2 Results of an interview with the Managing Director

Present: Managing Director
           RJH's Senior Analyst
           Senior Analyst's bag carrier

Date: 12 February 1990

Notes:
RCT was set up about fifteen years ago when the MD and a colleague (now Manager Public Courses) in a training company decided that they could do better. The company has grown over the years to fifty staff.

The original vision was to provide courses that not only provided the training needs of the customer but were also interesting, cost effective, and inspired the students to recommend the company to their colleagues.

Because of this, and especially due to the hard work of the initial ten within the company, the business has grown. RCT has a reputation for 'keeping the customer happy'. However, as the company has grown, the original vision has been diluted, and the MD wants to make sure that the company moves successfully through the next phase of its development. To this end, RCT has retained two external consultancies; one (ourselves) to provide the MD with a structured Business Information Technology (IT) Strategy, the other to use the business strategy to help the MD to implement the necessary reorganisation (note: we are to liaise with Mr Rupert Plantation of Change Management Ltd).

The MD's view of success to date is as follows:

1. Instructors are enthusiastic, extrovert, knowledgeable on the theory and have lots of real, practical experience.
2. The courses have the right mixture of lectures, 'homework', exercises and workshops.
3. The venues are always large enough to allow a relaxed but busy working environment.
4. The courses are demanding, requiring hard work — but students appear to 'enjoy' them.

On instructors

RCT employs instructors on the basis that they will, in general:

1. Help the production of new courses both in-house and for a specific client.
2. Give courses throughout the country.
3. Are willing to perform consultancy work.

In the last case, RCT does not want to hire out its consultants in a general and *ad hoc* manner (body-shopping); by offering consultancy to our existing or potential customers for training, RCT can offer a more complete customer service.

On public courses

Public courses are RCT's flagship operation. Hence, much effort is put into them in order to ensure the high quality of both the materials and the presentation. Furthermore, the courses must not be cancelled unnecessarily; the MD would rather spend some of his time drumming up extra customers. However, the decision to cancel must be made in a timely manner.

On types of course

RCT's goal is to offer a range of courses such that:

1. The skills of the instructor are fully utilised (in fact, the instructors stay happier if they are stretched).
2. The market need is satisfied.
3. Each instructor spends a reasonable amount of time (not sure of the proportions at this stage) on training, preparing courses and performing consultancy.

On money

RCT does not have a clear picture of the amount of money each particular course costs. Clearly, RCT needs to balance the various costs and income items, for example:

1. Preparing client-specific courses — would like to break even but depends on use to other courses.
2. Giving client-specific courses — reasonable positive income.
3. Preparing public courses — will be high-cost, unfortunately.
4. Giving public courses — must make a lot of money.
5. Performing consultancy — must at least break even.

RCT can ensure high positive income only from 'giving public courses'; the others are performed only to keep the instructors and customers 'happy'.

The MD is still deciding on RCT's financial future. The MD was hoping to improve profits by 30 per cent each year for three years, and then float on the full Stock Exchange. Recent market dealings (and shocks) has caused a rethink. The need now is to maximise income, minimise the cost of producing public courses, and ensure that no external work makes a loss. The increased 'profits' can then be used:

1. To provide a bonus for the staff, thus ensuring the retention of the high-quality staff that RCT require.
2. To develop new public courses.
3. To perform research into new techniques (such as HyperText, optical disks, computer-aided training (CAT)).
4. To build a cash mountain to take advantage of the current high interest rates and so tide RCT over lean times.

10.3 RCT business objectives

1. To provide high quality courses.
2. To provide a suitable environment for qualified and experienced staff.
3. To offer a comprehensive, market-driven set of courses.
4. To ensure high availability of courses; but be customer-orientated if a course does have to be cancelled.
5. To minimise costs in order to ensure suitable profits for profit-share schemes, developing new courses, training for the trainers and research into new training techniques.

10.4 The things that need measuring and how to measure them

1. To provide high quality courses.
  1.1 Customer satisfaction
    - Student course review ratings
    - Repeat business from client
    - Periodic customer reviews
    - Consultancy work from client
2. To provide a suitable environment for qualified and experienced staff.
  2.1 Staff turnover
    - Number of staff who leave by reason for leaving
    - Number of enquiries to join the company
  2.2 Staff enthusiasm
    - Number of courses attended by instructors
    - Mix of instructors giving courses (by location)
    - Student course review return (view of instructor)
  2.3 External reputation of staff
    - Papers presented at conferences
    - Qualifications held by staff
    - Invitations to seminars and the like
3. To offer a comprehensive, market-driven set of courses.
  3.1 Range of training available
    - Number of different courses
    - Number of courses not given by competitors
    - Number of courses offered by competitors and not by RCT
  3.2 Course attendance figures
    - Number of students on courses
4. To ensure high availability of courses; but be customer-orientated if a course has to be cancelled.
  4.1 Low level of possible bookings lost
    - Number of unsatisfied requests for a course place
    - Number of courses cancelled with students booked
  4.2 Requests for courses not available
    - Number of enquiries for courses not offered
5. To minimise costs.
  5.1 Profitability of the business
    - Return on capital investment
    - Actual amount of profit
    - Profit as amount per head of staff
  5.2 Course preparation
    - Cost of preparing courses
    - Number of common modules between courses
  5.3 Development spending
    - Amount available to profit share scheme

- Amount available for development of new courses
- Amount available for research into technical options

10.5 Targets for the measurable things

1.1 Customer satisfaction
- Student course review ratings
  AVERAGE OVERALL MARK 80 PER CENT
- Repeat business from client
  EVERY STUDENT FOR A CLIENT SHOULD GENERATE TWO MORE
  WITHIN TWO MONTHS
- Periodic customer reviews
  MD SHOULD SEE EVERY CLIENT, WHO HAS SENT 10 STUDENTS OR
  MORE, ONCE A MONTH
- Consultancy work from client
  30 PER CENT OF CLIENTS, WHO HAVE SENT 10 STUDENTS OR MORE,
  SHOULD BE OFFERING RCT 40 MAN-DAYS OF CONSULTANCY WORK
  A YEAR

2.1 Staff turnover
- Number of staff who leave by reason for leaving
  MAXIMUM OF 1 PER YEAR BECAUSE DISLIKE COMPANY
  MAXIMUM OF 0 PER YEAR BECAUSE OF BETTER OFFER
  MAXIMUM OF 2 PER YEAR, PERSONAL REASONS
- Number of enquiries to join the company
  1 PER YEAR INITIATED BY STAFF
  1 PER MONTH BECAUSE OF OUR REPUTATION

2.2 Staff enthusiasm
- Number of courses attended by instructors
  ALL COURSES TO HAVE AT LEAST 4 INSTRUCTORS
  80 PER CENT OF INSTRUCTORS WILLING TO GIVE ALL COURSES
- Mix of instructors giving courses (by location)
  EACH INSTRUCTOR ONLY GIVES SAME COURSE MAX. 4 TIMES PER
  YEAR
  EACH INSTRUCTOR VISITS EACH LOCATION AT LEAST ONCE A YEAR
  EACH INSTRUCTOR DOES AT LEAST 2 AWAY COURSES A YEAR
- Student course review return (view of instructor)
  EACH INSTRUCTOR TO ACHIEVE 75 PER CENT RATING
  NO INSTRUCTOR TO OBTAIN A SINGLE SCORE OF LESS THAN 50 PER
  CENT

2.3 External reputation of staff
- Papers presented at conferences
  20 PER CENT OF STAFF TO PRESENT 1 PAPER A YEAR
- Qualifications held by staff
  70 PER CENT OF INSTRUCTORS TO HAVE A FIRST DEGREE
  5 PER CENT OF INSTRUCTORS TO BE GAINING A FURTHER DEGREE

- Invitations to seminars and the like
  10 PER CENT OF INSTRUCTORS ARE INVITED EACH YEAR

3.1 Range of training available
- Number of different courses
  AIMING FOR 40 DIFFERENT COURSES BY END OF 2000
  10 PER CENT MORE COURSES THAN 'SMART-COURSES INC.'
- Number of courses not given by competitors
  10 PER CENT OF COURSES NOT GIVEN BY ANYONE ELSE
- Number of courses offered by competitors and not by RCT
  10 PER CENT

3.2 Course attendance figures
- Number of students on courses
  EACH COURSE TO BE AT LEAST 80 PER CENT FULL

4.1 Low level of possible bookings lost
- Number of unsatisfied requests for a course place
  NO GREATER THAN 5 PER COURSE
- Number of courses cancelled with students booked
  NO GREATER THAN 5 PER YEAR

4.2 Requests for courses not available
- Number of enquiries for courses not offered
  ZERO IF REQUESTED COURSE IS OFFERED ELSEWHERE

5.1 Profitability of the business
- Return on capital investment
  LOOKING FOR 25 PER CENT FOR NEXT THREE YEARS
  RISING TO 33 PER CENT AFTER THAT
- Actual amount of profit
  To be determined (TBD)
- Profit as amount per head of staff
  TBD

5.2 Course preparation
- Cost of preparing courses
  TBD
- Number of common modules between courses
  TBD

5.3 Development spending
- Amount available to profit share scheme
  10 PER CENT
- Amount available for development of new courses
  10 PER CENT
- Amount available for research into technical options
  10 PER CENT

10.6   Gathering candidate Entity Types

10.6.1   Stage 1. Read information given and list entities

The following entities were found to be of relevance after reading the information in the previous sections:

Public training course
Customer-specific training course
Client's premises
Hotel
Room (training)
Computer equipment
Client
Course module
Course
Tutor material
Student material
Course material
Tutor
Student
Computer software
Cost information
Income (from various sources)
Presentation material (of course material)
Exercise material (of course material)
Overhead projector slides (of presentation material)
Question
Answer
Tutor explanatory notes
Course duration
Venue
Equipment
Costs (associated with various items)
Public course schedule
Instructor's ability
Booking information
Attendance
Accommodation/meals/refreshments
Course review forms
Course administrator

10.6.2   Stage 2. List candidate Entity Types

'Courses' are run as 'Public' or 'Customer-specific'
'Course' is held in a 'Room' at 'Client premises' or 'Hotel/Venue'

'Course Material' comprises: 'Computer Equipment'
                                'Tutor Material/Explantory Notes'
                                'Student Material'
                                'Computer Software'
                                'Overhead Projector Slides'
                                'Questions'
                                'Answers'

'Client'
'Course Module'
'Tutor/Instructor'
'Student'
'Costs'
'Income'
'Public Course Schedule'
'Instructor's Ability'

10.6.3   Stage 3. First-pass Data Model

At a brief meeting (on such-and-such date) it was decided to leave out the costs/
income aspects at this level. The first-pass Data Model is shown in Figure 10.2.

Each 'Course Module' is associated with 'Course Material'
Each 'Course' is associated with 'Course Modules'
Each 'Course' is associated with 'Course Offering'
Each 'Course Offering' may be 'Public' or 'Customer-specific'
Each 'Instructor' is associated with 'Courses'

Figure 10.2
The first-pass Data
Model.

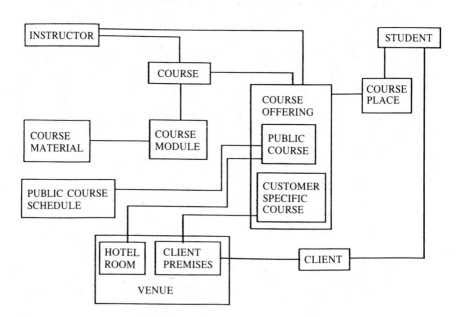

Each 'Course Offering' may have a (qualified) 'Instructor' booked for
it (*we know this* and so state it)

Each 'Course Offering' is associated with 'Course Places'

Each 'Student' is associated with 'Course Places'

There is a relationship between a client-specific 'Venue' and a 'Client'

Each 'Client' is associated with 'Students'.

10.7   High-level Data Model

After a review of the first-pass Data Model and more discussions, the high-level
Data Model was agreed. This is shown in Figure 10.3.

The following is the supporting narrative for the high-level Data Model for
RCT (Version 1.3, 21 February 1990):

Each 'Course' is split into one or more 'Course Modules'

Each 'Course Module' is made up from one 'Module'

Each 'Module' may be used as one or more 'Course Modules'

Each 'Module' includes one or more types of 'Course Material'

Each 'Course may be offered as one or more 'Course Offerings'

The Entity Type 'Course Offering' has subtypes 'Public Course' and
'On-site Course'

Figure 10.3
High-level Data
Model for RCT.

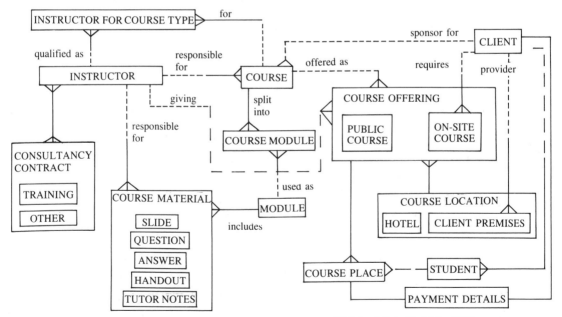

High-level Data Model for RCT
Version 1.3, 21 February 90, RJH
Status: Agreed at feedback presentation

Each 'Instructor' may be qualified as one or more 'Instructors for Course Types'

Each 'Instructor' may be responsible for one or more 'Courses'

Each 'Instructor' (may be involved) in one or more 'Consultancy Contracts'

Each 'Consultancy Contract' (may be serviced) by one or more 'Instructors'

(TECHNICAL QUERY: Is this correct?)

Each 'Course' may have one or more 'Instructors for Course Type' qualified to give the course

Each 'Course Offering' may have a (qualified) 'Instructor' booked to give it

Each 'Instructor' may be giving one or more 'Course Offerings'

Each 'Course Offering' is made up of one or more 'Course Places'

Each 'Student' may be booked on one or more 'Course Places'

Each 'Course Place' may have one 'Student' booked on it

Each 'Client' may have one or more 'Students'

Each 'Client' may be the sponsor for one or more 'Courses'

Each 'Client' may require one or more 'On-site Courses'

The Entity Type 'Course Location' has subtypes 'Hotel' and 'Client Premises'

Each 'Client' may be the provider of one or more 'Client Premises'

Each 'Course Offering' is held at one 'Course Location'

Each 'Course Location' holds one or more 'Course Offerings'

TECHNICAL QUERY: Does the business want to hold details on course locations before course offerings are assigned to them? If so, the relationships shown on the diagram will be wrong.

TECHNICAL QUERY: What are the relationships between 'Client', 'Course Place' and 'Payment Details'?

10.8   High-level Activity Hierarchy Diagram

10.8.1   *Outline Activity Hierarchy Diagram*

The outline Activity Hierarchy Diagram is shown in Figure 10.4.

10.8.2   *Description of high-level activities for RCT*

A. Maintain company strategy

Ensure that the company business plan always contains a long-term element, and that business priorities are set according to an acceptable balance of short- and long-term considerations.

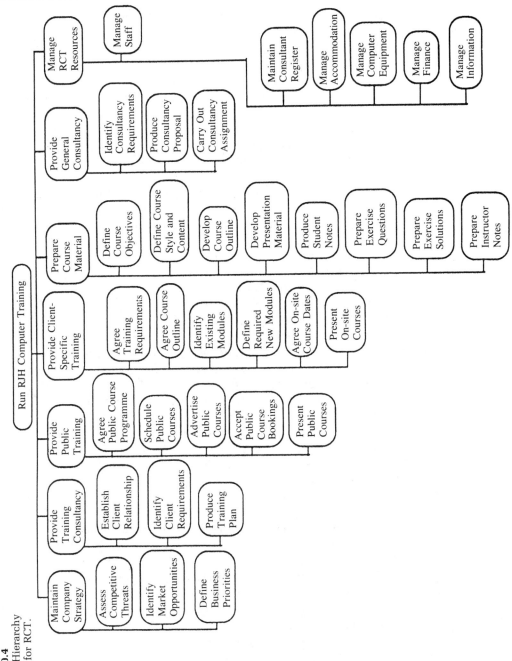

Figure 10.4
Activity Hierarchy
Diagram for RCT.

A1   Assess competitive threats
A2   Identify market opportunities
A3   Define business priorities

B.   Provide training consultancy

Provide advice to clients on how best to meet the training needs of their staff. This is generally done at no charge, unless an unusually large amount of work is required.

B1   Establish client relationship
B2   Identify client requirements
B3   Produce training plan

C.   Provide public training

Provide well-advertised training courses to the general public on a regular basis. Though only moderately profitable in its own right, contacts made in this way are the main source of RCT's other business.

C1   Agree public course programme
C2   Schedule public courses
C3   Advertise public courses
C4   Accept public course bookings
C5   Present public courses

D.   Provide client-specific training

Provide training courses which are not (yet) intended for the general public, and whose costs must therefore be justified on the courses' value to a specific sponsoring client. The courses may use existing material, new material or a mixture of the two.

D1   Agree training requirements
D2   Agree course outline
D3   Identify existing modules
D4   Define required new modules
D5   Agree on-site course dates
D6   Present on-site courses

E.   Prepare course material

Design and produce the necessary materials for a new course or module.

E1   Define course objectives
E2   Define course style and content
E3   Develop course outline
E4   Develop presentation material
E5   Produce student notes
E6   Prepare exercise questions
E7   Prepare exercise solutions
E8   Prepare instructor notes

F.   Provide general consultancy

Provide general consultancy services in those areas in which RCT staff have expertise. While not the principal area of RCT's business, this is encouraged as it provides opportunities for staff to broaden their experience and obtain potential case studies/examples for course material. It also deepens RCT's relationships with clients.

F1   Identify consultancy requirements
F2   Produce consultancy proposal
F3   Carry out consultancy assignment

G.   Manage RCT resources

Manage the availability of the various resources without which RCT could not function — staff, equipment, finance, etc.

G1   Manage staff
G2   Maintain consultant register
G3   Manage accommodation
G4   Manage computer equipment
G5   Manage finances
G6   Manage information

10.9   Mapping matrices

10.9.1   Activity versus Entity Type

Figure 10.5 shows the CRUD matrix, mapping the Entity Types on the high-level Data Model (Figure 10.3) against the top-level activities shown on the Activity Hierarchy Diagram (Figure 10.4). The table shows information for the top-level activities plus 'provide public training' broken down in more detail.

Figure 10.5
High-level CRUD matrix.

C=CREATE, R=READ U=UPDATE, D=DELETE	Instructor	Course	Client	Course Offering	Course Module	Course Place	Student	Payment Details	Consultancy Contract	Course Material
1. Maintain Company Strategy	R	C	R	R						
2. Provide Training Consultancy			U						C	
3. Provide Client Specific Training		C	U	C	C	C				
4. Prepare Course Material		R			R					C
5. Provide General Consultancy									C	
6. Manage RCT Resources	C									
7. Provide Public Training										
7.1 Agree Public Course Programme				C						
7.2 Schedule Public Courses	R	R				C				
7.3 Advertise Public Courses				R						
7.4 Accept Public Course Booking			U	R		R	C	C		
7.5 Present Public Courses				U	R	R	U	U		

TECHNICAL QUERY: Why are there no deletes?
BUSINESS ISSUE: When should historical data be removed from the system?

10.9.2 Current organisation

Figure 10.6 shows the current organisation hierarchy within RCT.

Figure 10.6
RCT organisation
hierarchy.

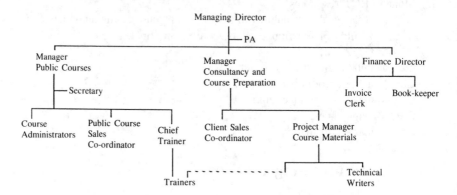

10.9.3 Organisation versus activity

Figure 10.7
The organisation
versus activity tick
matrix.

Figure 10.7 shows the tick matrix showing what parts of the RCT organisation use and/or are involved with the top-level activities. Again, the top-level activities are shown together with 'provide public training' broken down.

	MD	FD	Manager public courses	Course admin.	Public courses sales	Chief trainer	Trainer	Manager consultancy course prep.	Client sales	PM course materials
Maintain Company Strategy	✓		✓					✓		
Provide Training Consultancy						✓	✓	✓	✓	
Provide Client-Specific Training						✓	✓	✓	✓	
Prepare Course Material						✓	✓	✓		✓
Provide General Consultancy						✓	✓	✓		
Manage RCT Resources		✓								
Provide Public Training										
Agree Public Course Programme	✓	✓				✓				
Schedule Public Courses		✓								
Advertise Public Courses					✓					
Accept Public Course Booking					✓					
Present Public Courses				✓			✓			

10.10   Implementation priorities

Even as the Requirements Analysis was progressing, it was clear that the highest priority was to enhance RCT's handling of public courses.

It was decided to revisit the analysis to consider further priorities at a later date. With respect to 'public courses', the key was to schedule the courses and then ensure adequate bookings and the availability of a suitable trainer. When a public course has to be cancelled, for whatever reason, there needs to be a slick way of rebooking any booked students.

Hence a plan would be put forward further to analyse 'Schedule Public Courses', which is a subactivity of 'Provide Public Training'.

10.11   Management agreement for 'Schedule Public Courses'

The outline plan to do the following was agreed:

> Interview the Manager Public Courses
> Expand the Data Model
> Produce detailed Activity Hierarchy Diagram
> Produce Data Flow Diagrams

The plan will be formally presented at the start of the Analysis Phase.

11 Case study: Activity and Data Analysis Phase for 'Schedule Public Courses'

11.1 Plan

In order to produce the detailed analysis of 'Schedule Public Courses', it was deemed important to follow through the Business Analysis of RCT and perform some high-level analysis of 'Provide Public Training'. This analysis would then be a sound basis for moving into the area of 'Schedule Public Courses'.

Once the analysis of 'Schedule Public Courses' is complete, a Project Meeting will be held so that one particular area can be identified for more detailed analysis with a view of progressing that area into design and implementation. The agreed plan is shown in Figure 11.1.

Figure 11.1
The agreed project plan for the Activity and Data Phase.

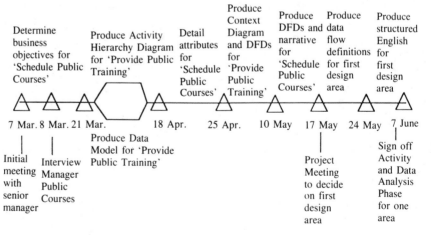

Activity and Data Analysis Project Plan
5 March 1990, RJH
Status: As agreed

11.2   Interview with Manager Public Courses

Present: Manager Public Courses
           RJH's Senior Analyst

Date: 8 March 1990

Notes:

Bill, the Public Course Manager, joined as an instructor ten years ago, but has had management training and experience. He now spends most of his time in administration and customer visits, but still teaches occasionally.

The main problem appears to be in performing the balancing act between turning customers away from over-booked courses and having to cancel courses due to lack of numbers. It is impossible to predict the level of bookings from course to course. Many bookings come in the last four weeks when it is really too late to cancel and also too late to change to a larger venue. RCT do have the right under their standard agreement to cancel at only two weeks' notice, but rarely do so as that would lose goodwill. At the present level of charges, RCT require about six students to break even on a course, though the exact figure depends on the venue and the type of course.

11.3   Business objectives, measures and targets for 'Schedule Public Courses'

1. To maximise profit on courses.
    1.1 Profitability
         - Gross profit on courses
           £80,000
         - Profit as amount per actual student day
           £40
         - Profit as amount per course place offered
           £100
         - Average number of students per course
           MINIMUM 5, AVERAGE DEPENDENT ON COURSE
2. To offer a balanced range of courses over a period.
    2.1 Range of training available
         - Number of different courses offered over a period
           TBD
         - Number of courses not offered by competitors
           TBD
         - Number of courses offered by competitors but not RCT
           TBD
3. To ensure well-qualified instructors on every course.
    3.1 Enough qualified instructors for each course
         - Number of courses cancelled due to lack of instructor
           0

- Number of courses not offered due to a lack of an instructor
  0
- Number of instructors available for each course type
  3

4. To achieve a high level of customer satisfaction.
   4.1 Customer satisfaction
   - Student course review ratings for course
     90 PER CENT
   - Student course review ratings for trainer
     90 PER CENT
   - Student course review ratings for location
     75 PER CENT

5. To avoid losing business due to non-availability of courses.
   5.1 Low level of possible bookings lost
   - Number of unsatisfied requests for public course places
     10 PER MONTH
   - Number of enquiries for course not offered
     15 PER MONTH
   - Number of students rebooked due to cancelled courses
     5 PER MONTH
   - Number of students who cancel after being rebooked due to cancelled course
     0

11.4 Things that help the objectives to succeed

1. Profitability.
   - Low break-even levels
     Low development costs
     Low running costs
   - High attendance
     Effective advertising
     Effective training guidance service
2. Range of training available.
   - Breadth (range of subjects)
     Technically-aware staff
     Size of research budget
     Collaborative arrangements
   - Depth (from introductory to advanced)
     Experienced staff
     Size of development budget
     Size of customer base
3. Enough qualified instructors for each course.
   - High level of staff experience
     Careful recruiting

Opportunities for continuing experience
Low staff turnover

- Minimum level of experience required
  Comprehensive course materials prepared in advance
  Guidance notes for instructors
  Continuous updating of course material
  Access to more experienced instructor when necessary
- Effective introductory plan for each new course
  Thorough review after first few presentations
  Schedule instructor induction into each course

4. Customer satisfaction.

- General customer image
  Effective telephone answering
  Clear and prompt correspondence
  Accurate administration
- Image of instructors
  Technical competence
  Presentation skills
- Image given by course material
  Comprehensive material
  Well-produced material
  Correction of errors after each course
- Quality of course content
  Clear definition of intended audience
  Clear definition of objectives
  Adequate preparation plan
  Prepared by knowledgeable staff
  Thorough review process
- Quality of venues
  Large rooms
  Smoking/no-smoking areas
  Lunch and break facilities
  Overnight accommodation if residential

5. Minimum level of possible bookings lost.

- Courses available in requested subjects
  Market research in anticipated trends
  Staff recruitment to fill gaps
- Courses available at required depth
  Keep experienced staff
  Use external consultants
- Public courses available when required
  Responsive scheduling
  Reduced lead time in rescheduling

11.5   Things that constrain the success of the objectives

1.   Must use only existing computer equipment.
2.   Must maintain existing format of customer data, or provide conversion.
3.   Cannot change prices once brochure is printed.
4.   Must give adequate notice to hotels (four to six weeks).
5.   Must give adequate notice of cancellation to customers (four weeks) and must then offer rebooking within one week.
6.   Must not schedule same instructor away from base on three consecutive weeks, and not on two consecutive weeks unless in same location.
7.   Some courses have prerequisites, which must be available at similar frequency.

11.6   Problems

1.   Customers upset by late cancellation.
2.   High cost of running under-subscribed courses.
3.   Difficult to know likely demand in advance.
4.   Have to deal with sudden instructor absence.
5.   Hotel accommodation is sometimes difficult to find.
6.   Difficult to predict requirements for specialised equipment.
7.   Too few qualified instructors for some courses.

11.7   Activity Hierarchy Diagram for 'Provide Public Training'

11.7.1   Activity Hierarchy Diagram

The Activity Hierarchy Diagram for 'Provide Public Training' is shown in Figure 11.2.

11.7.2   Description of activities for 'Provide Public Training'

In alphabetical order:

Accept booking change
   Record a change of student details for a course booking. This does not cover a change of date, which is treated as a cancellation followed by a new booking.
Accept booking confirmation
   Record student details which are not provided when a provisional booking was made and initiate the invoicing process.
Accept cancellation
   Record the cancellation of a booked course place and initiate a partial credit/refund.

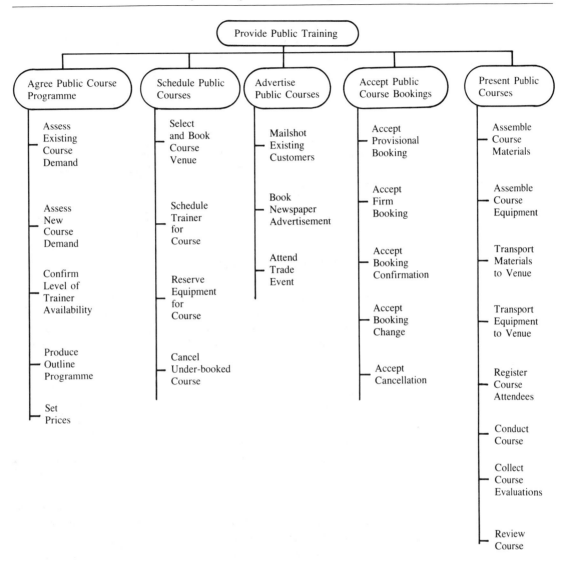

Figure 11.2
Activity Hierarchy
Diagram for 'Provide
Public Training'.

Accept firm booking
Record details of a firm booking for one or more course places, including student details.

Accept provisional booking
Record the details of a course place booking for which no student details are yet available. A provisional booking is held until two weeks before the start of the course and there is no penalty for cancellation.

Accept public course bookings
Provide a booking service for public courses by telephone, mail or other contact. This includes cancellations.

Advertise public courses
> Promote, by advertising, direct mail, personal contact or other means, RCT's programme of public courses.

Agree public course programme
> From consideration of past booking levels, known future demand, expected market trends and other factors, decide which public courses to offer, and at what frequency, in the period being planned.

Assemble course equipment
> Gather together all equipment to be provided by RCT for a course, ensuring that all is in working order and complete with cables and other portable components.

Assemble course materials
> Assemble the handouts, slides and any other materials for students and trainer required for a course, ensuring that all are complete and up-to-date.

Assess existing course demand
> Assess the public demand for places on existing courses, and hence the number and desired location of courses to be run.

Assess new course demand
> Assess the demand for courses which have not previously been covered in the public course programme. These will usually be courses which have been developed to meet a known requirement for specific clients. The result will be an estimated number of places, and hence courses to be provided.

Attend trade event
> Attend a show, conference or seminar to promote RCT's services, sometimes in general and sometimes with an emphasis on particular courses. This may include speaking on training and also writing papers for conference proceedings.

Book newspaper advertisement
> Arrange for an advertisement to appear in a newspaper, magazine or other appropriate publication. This includes the preparation of any necessary material.

Cancel under-booked course
> When the income from a course appears likely to be unacceptably low, cancel the course and make alternative arrangements for any places already booked on it.

Collect course evaluations
> Ensure that all students have completed their course evaluations and collect them for subsequent evaluation.

Conduct course
> Carry out the presentation of a course, noting any problems for subsequent resolution.

Confirm level of trainer availability
> Check that the level of trainer availability will be broadly adequate

for the planned quantity of course offerings

Mailshot existing customers

Send course brochures to existing customers, both regularly and to promote specific courses.

Present public courses

Present course at premises arranged by RCT and promoted as part of the public course programme.

Produce outline programme

Produce an outline programme of public courses giving an indication of the numbers and locations of each course type and preferred dates. This will be subject to considerable change during the scheduling process.

Register course attendees

At the start of a course, register students as they arrive and ensure that any problems of accommodation or other facilities are resolved. Produce a student list for the trainer and charging details, including any absences, substitutions and known reasons for absence, for use in producing credit notes or invoice adjustments.

Reserve equipment for course

Reserve an item of computer or other equipment for a specific course.

Review course

Review the course evaluation forms, and any feedback provided by the trainer, and note any possible improvements for future action.

Schedule public courses

Fix dates for the planned public courses, taking into account availability of rooms, instructors and, sometimes, equipment.

Schedule trainer for course

Select a qualified trainer for a course. There are restrictions on the allocation of trainers to courses away from base and on the number of consecutive weeks a trainer may be training.

Select and book course venue

Given the nature of a planned public course, select an appropriate venue and make a provisional booking. Factors taken into account include the need for syndicate rooms or computer equipment, as well as seasonal factors for some venues, and whether the course is residential.

Set prices

For each course to be offered in a public course programme, set the price to be charged for a place at each offering of that course. Even if the course is not new, the price is always reviewed.

Transport equipment to venue

Transport the required equipment to the course venue and ensure that it is assembled and in working order.

Transport materials to venue

Transport the required course materials to venue.

11.8   Data Model extract for 'Provide Public Training'

The Data Model shown in Figure 11.3 has been extracted and further refined from the high-level Data Model of RCT as agreed in the Business Analysis Phase. The baseline high-level Data Model was presented in Chapter 10, Figure 10.3.

The key refinements from the baseline are:

1. Relationships have been labelled in both directions.
2. The relationship between 'Instructor' and 'Course Offering' has been made 'may' in both directions. A course offering, when set up, will not have an instructor associated with it: this is why this piece of work is being carried out — how to schedule an instructor to a course offering.
3. Extra detail has been added around 'Instructor'; now an instructor has one or more time commitments. Time commitments are chunks of time when the instructor is unavailable to give a course, such as holidays and doing consultancy work. It does *not* include time giving courses, as this is logged via the relationship with course offering.

Figure 11.3
Data Model extract for 'Provide Public Training'.

4. The relationship between a course place and the client and student has been refined. A client can reserve a course place without naming the student. When the student's details are known, the student/course place relationship is made

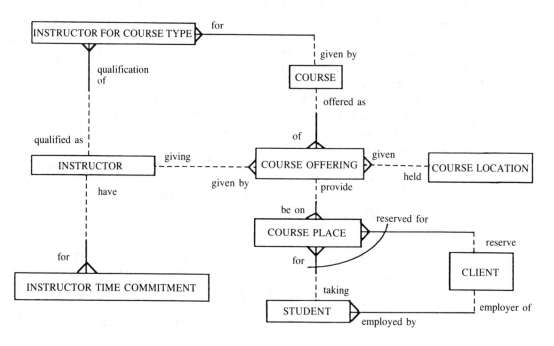

Data Model for 'Provide Public Training'
Version 1.1, RJH
Status: For review

and kept, and the client/course place is cancelled. We still know who the client is if we wish to find out, via the student/client link.

11.9   Description of the Data Model extract

The following is the narrative associated with the Data Model shown in Figure 11.3.

Each 'Course' may be offered as one or more 'Course Offerings'
Each 'Course Offering' must be of one 'Course'
Each 'Course Offering' may provide one or more 'Course Places'
Each 'Course Place' must be on one 'Course Offering'
Each 'Course Place' must be either reserved for one 'Client' or for one 'Student', but not both
Each 'Client' may reserve one or more 'Course Places'
Each 'Student' may be taking one or more 'Course Places'
Each 'Client' may be the employer of one or more 'Students'
Each 'Student' must be employed by one 'Client'
Each 'Instructor' may be qualified as one or more 'Instructor for Course Type'
Each 'Instructor for Course Type' must be a qualification of one 'Instructor'
Each 'Course' may be given by one or more 'Instructors for Course Type'
Each 'Instructor for Course Type' must be for one 'Course'

The above four relationships define the facts that:

Each 'Course' can be given by one or more 'Instructors' (provided that the instructors are qualified to do so)
Each 'Instructor' can give one or more 'Courses' (provided that he or she is qualified to do so)
Each 'Instructor' may be giving one or more 'Course Offerings'
Each 'Course Offering' may be given by one 'Instructor'
Each 'Instructor' may have one or more 'Instructor Time Commitments'
Each 'Instructor Time Commitment' must be for one 'Instructor'
Each 'Course Offering' may be given at one 'Course Location'
Each 'Course Location' may hold one or more 'Course Offerings'

11.10   Attributes of the Data Model relevant to 'Schedule Public Courses'

The following is a list of attributes for each Entity Type shown in the Data Model. The italicised attributes make up the identifier for the Entity Type.

ENTITY TYPE	ATTRIBUTE	VALUE
Instructor	*Instructor id*	3 chars
	Surname	15 chars
	First name	8 chars
Instructor for Course Type	*Instructor id*	
	Course id	10 chars
	Number of times given course	1–999
	Owner of course?	Y/N
	Helped to write it?	Y/N
Instructor Time Commitment	*Instructor id*	
	Commitment number	1–999999
	Week number	1–999
Course	*Course id*	
	Course name	40 chars
	Number of weeks to book	1, 2 or 3
	Number of days	1–10
	Maximum number of places	1–99
	Equipment required	70 chars
Course Offering	*Course id*	
	Sequence number	1–999
	Week number	
	Start day	0–4 (Mon–Fri)
	Status	<5 no instructor
	Geographic code	0–99
	Instructor id	
	Venue id	
	Equipment status	5 requested
		9 booked
Course Place	*Course id*	
	Sequence number	
	Place number	1–99
	Status	0–9
	Reserved by (ref. to client)	
	Student booked (ref. to student)	
	Student registered (ref. to student)	

Student	*Student id*	1–999
	Student surname	15 chars
	Comments	70 chars
	Client id	
Client	*Client id*	1–999
	Client name	20 chars
	Client address	70 chars
	Comments	70 chars
Course Location	*Venue id*	1–99
	Reservation address	70 chars
	Geographic code	
	Maximum number of students	1–99

11.11 Context Diagram for 'Provide Public Training'

The Context Diagram for 'Provide Public Training' is shown in Figure 11.4.

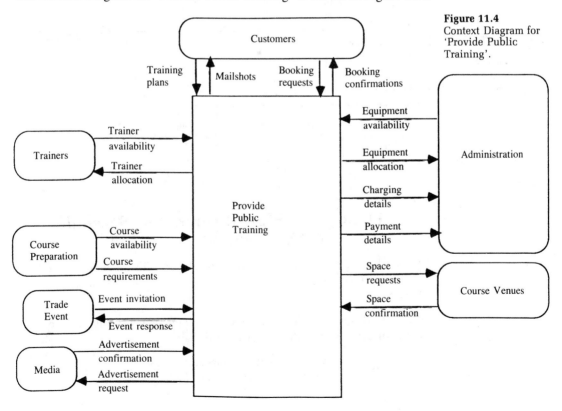

Figure 11.4
Context Diagram for 'Provide Public Training'.

11.12   Data Flow Diagram for 'Provide Public Training'

The Data Flow Diagram for 'Provide Public Training' is shown in Figure 11.5.

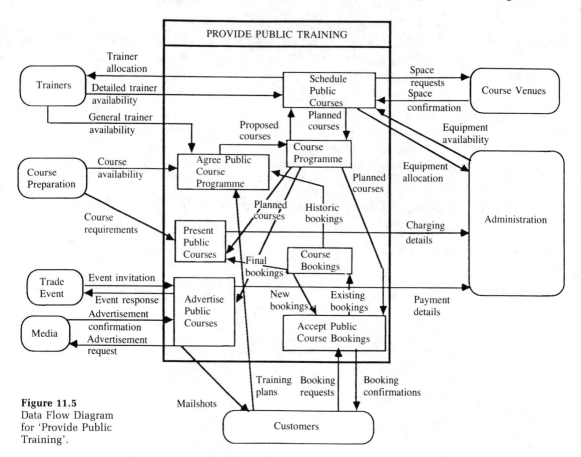

Figure 11.5
Data Flow Diagram
for 'Provide Public
Training'.

11.13   Data Flow Diagram for 'Schedule Public Courses'

The Data Flow Diagram for 'Schedule Public Courses' is shown in Figure 11.6.
    The analyst has logged a technical query against this Data Flow Diagram, as follows:

TECHNICAL QUERY: The Data Flow Diagram shown in Figure 11.6 has a number of data flows with the same name, namely:

                        Flows 42 and 43: Placed courses
                        Flows 26 and 46: Planned courses
                        Flows 44 and 45: Staffed courses

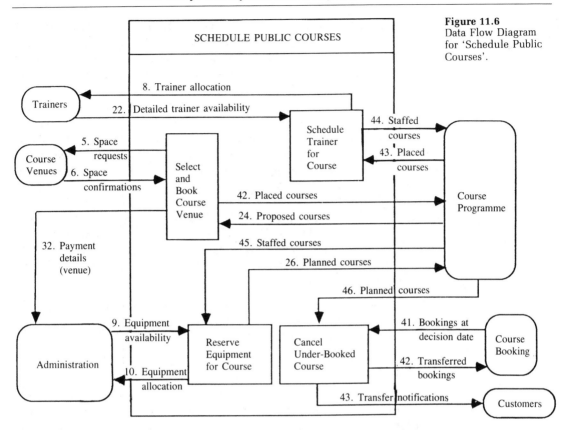

Figure 11.6
Data Flow Diagram
for 'Schedule Public
Courses'.

Are the contents of the flows with the same name exactly the same? (Whatever the answer, I strongly believe that flows should have slightly different names so that there is absolutely no confusion — but it all depends on the circumstances and the audience.)

11.14 Data Flow Diagram narrative for 'Schedule Public Courses'

The following is the textual description of the Data Flow Diagram for 'Schedule Public Courses', which is shown in Figure 11.6.

The activity 'Schedule Public Courses' has four lower-level activities, namely:

1. Select and book course venue.
2. Schedule trainer for course.
3. Reserve equipment for course.
4. Cancel under-booked course.

The activity 'Select and Book Course Venue' requires information on the proposed courses (flow 24 from 'Course Programme'). Requests for adequate accommodation at course venues are made (flow 5 'Space requests' to 'Course

Venues'), and then options, and subsequently confirmations, returned (flow 6 'Space confirmations' from 'Course Venues'). Once the activity has agreed the location and cost associated with the location, the course programme is updated (flow 42 'Placed courses') and the payment details with respect to the venue sent to Administration (flow 32 'Payment details (venue)').

In order to schedule a trainer for the course, those courses which now have a venue but still do not have a trainer assigned to them (flow 43 'Placed courses' from 'Course Programme') are considered so that a trainer can be found for them. Information on the availability of trainers is obtained (flow 22 'Detailed trainer availability') and decisions made as to which trainers will be used on which courses. The results are reported back to the trainer information (flow 8 'Trainer allocation') and the course programme (flow 44 'Staffed courses').

In order to reserve equipment for the course, those courses which have both a venue and a trainer allocated (flow 45 'Staffed courses') are considered. From a list of available equipment (flow 9 'Equipment availability'), equipment is matched to the requirements and allocated via the administration function (flow 10 'Equipment allocation').

TECHNICAL QUERY: Must ensure that I understand what happens if equipment is not available. Similarly for venue and trainer unavailability. It may be necessary to do an Entity Life History on 'Course'.

The public course details with trainer, venue and equipment is now logged in the course programme (flow 26 'Planned courses').

The activity 'Cancel Under-booked Course' reviews the planned courses (flow 46 'Planned courses') and checks the number of bookings at the cut-off date for cancelling an under-booked course (flow 41 'Bookings at decision date'). If a course is cancelled, those already booked on the course are, if possible, transferred to another course offering (flow 42 'Transferred bookings') and the customer formally notified of the change (flow 43 'Transfer notifications').

11.15   Notes on the Project Meeting to decide the first Design Area

The Data and Activity Analysis deliverables to date were presented and reviewed at the Project Meeting held on 17 May 1990. The meeting agreed that the trial implementation should be for 'schedule trainer for course' as this was the area causing most problems, especially as it is currently carried out manually.

The plan to the end of the Activity and Analysis Phase is summarised as follows:

17 May–24 May	Produce data flow definitions relevant to the design area 'Schedule Trainer for Course'.
24 May–6 June	Produce structured English for the activity 'Schedule Trainer for Course'.

7 June	Hold Project Meeting in order to:
	1. Perform Feedback Presentation of analysis work relating to 'Schedule Trainer for Course';
	2. Sign-off the deliverables if possible;
	3. To agree the design phase planning.

11.16   Data flow definitions for 'Schedule Trainer for Course'

44 Staffed courses

'Course Offering'
Course id
Sequence number
Status (set to 8)
Instructor id (set to chosen id)

45 Placed courses

'Course Offering'
Course id
Sequence number
Status (only interested if <5)
Week number
'Course'
Course id
Number of weeks to book

22 Detailed trainer
availability

'Instructor for Course Type'
Instructor id
Course id
'Instructor Time Commitment'
Instructor id
Commitment number
Week number
'Instructor'
Instructor id
Surname
First name

8 Trainer
allocation

'Instructor for Course Type'
Instructor id
Course id
Number of times given course (add 1)

'Instructor Time Commitment'
Instructor id
Commitment number (add 1 to it)
Week number

11.17   Structured English for 'Schedule Trainer for Course'

Having now produced a Data Model and some Data Flow Diagrams, it is important to ensure that:

1. The activities on the Data Flow Diagrams are well understood.
2. The activities can 'move around' the Data Model using the relationships between the Entity Types.

Structured English is a technique for achieving this. There are many different forms of structured English and the Project Manager should choose a suitable form that matched the needs of the project. If, for example, a 4GL-style language is being used for the implementation, then perhaps a subset of the language could be used for the structured English.

The following structured English definition for 'Schedule Trainer for Course' uses a simple technique which is described below.

In order to ensure that the activity can 'walk' the Data Model, there are two constructs: READ and FOR EACH. The format for READ is either:

READ Entity Type Name USING Attribute (for which we know a value);
    or
READ Entity Type Name USING Relationship Name Entity Type Name

READ reads one and only one 'instance' of an Entity Type based on the USING clause.

If the relationship is one-to-many and we wish to read all of the manys, then the form FOR EACH is used:

FOR EACH Entity Type Name USING Relationship Name Entity Type Name

Although this style is a bit pedantic, it does check our understanding of the structure of the model.

Once an Entity Type is gathered in, then some procedural text is required, quite often introduced by the construct IF. In this example, I have used plain English for the procedural text. The other useful construct used in this example is CONNECT, which states that a new relationship needs setting up.

As in all good structured languages, each IF and FOR is suitably terminated with an END IF and END FOR respectively. Lastly, lines starting with * are comments.

So, there now follows the structured English for the activity 'Schedule Trainer for Course':

ACTIVITY Schedule Trainer for Course
* May decide to search for all public courses with no trainers
* and/or look at specific courses, but assume that one is selected
* via COURSE-OFFERING-NUMBER
*
READ 'COURSE OFFERING' USING COURSE-OFFERING-
   NUMBER
READ 'COURSE' USING OF 'COURSE-OFFERING'
FOR EACH 'INSTRUCTOR-FOR-COURSE' USING FOR 'COURSE'
FOR EACH 'INSTRUCTOR' USING QUALIFIED-AS 'INSTRUCTOR-
   FOR COURSE'
* Check if instructor is already doing a course

FOR EACH 'COURSE-OFFERING' GIVEN-BY 'INSTRUCTOR'
   IF course dates of this course-offering clash with the course
   dates of the course identified by course-offering-number then
   reject this instructor and move onto the next instructor.
   END IF
END FOR

* If instructor not already doing a course check to see if he has
* other time commitments
*

FOR EACH 'INSTRUCTOR-TIME-COMMITMENT' OF this
   'INSTRUCTOR'
   IF dates of this commitment clash with the course dates of the
   course identified by course-offering-number then reject this
   instructor and move onto the next instructor.
   END IF
END FOR

* So if this instructor is not rejected we could use him/her
* One day it might be worth selecting all possible instructors
* and allowing the user to choose the most suitable *

   IF this 'INSTRUCTOR' is not rejected THEN
      CONNECT 'INSTRUCTOR' TO 'COURSE OFFERING' USING
         'given by'
      EXIT ACTIVITY
   END IF
END FOR
END FOR
IF no instructors have been found THEN output a suitable error/
   information message
END ACTIVITY

Part 4

Design and implementation:
general techniques

12 Moving from Analysis to the Design Phase

12.1 Introduction

The Data and Activity Models produced in the Activity and Data Analysis Phase are pivotal pieces of information. During the Analysis Phase, the analyst must ensure that the models can be understood by the users and managers, and yet are detailed enough to be fleshed out ready for the Design Phase.

The analyst starts the movement towards the Design Phase. No longer is there a need to worry too much about the users and managers understanding the models: it is now more important that the designer can pick up the specification and produce a detailed design for the automated support required. The analyst must first of all refine the Data Model to ensure that the model is complete, consistent and well understood. This process is described in section 12.2. The designer then takes the refined Data Model together with the activity definitions and produces a definition of the procedures that are needed to support the activities. The processes for doing this are described in the remaining sections of this chapter.

12.2 Refining the Data Model

12.2.1 Introduction

There comes a time in data modelling when the emphasis of the modelling is targeted towards design and not just presenting the ideas to senior managers and users. The final Data Model should be detailed enough for a designer to design the data structures and produce the database definition procedures. To this end, the Data Model must be complete and unambiguous.

Completeness includes:

    Complete description of all attributes
    Volume information
    Relationships named in both directions
    Cardinality defined as zero, one, one or more, etc.

The following briefly describes techniques to refine the model in order to make it generally clearer, unambiguous and suitable as the definition for the database design. The techniques described are:

Searching for redundant relationships
Modelling time
Historical records
Reviewing one-to-one relationships
Reviewing many-to-many relationships between different Entity Types
Reviewing many-to-many relationships between the same Entity Type
Reviewing integrity conditions

12.2.2   Redundancy

Let us take, for example, a model where there is a relationship between 'Client' and 'Type of Car' (buys), a relationship between 'Client' and 'Order' (places), and a relationship between 'Order' and 'Type of Car' (specifies). This model is shown in Figure 12.1.

Figure 12.1
Is there a redundant
relationship in this
diagram?

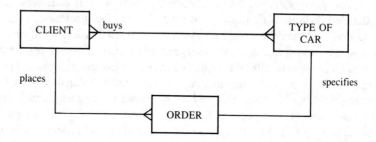

In this example, the relationship 'buys' needs further consideration as it may be redundant. If 'Order' always exists for as long as the relationship between 'Client' and 'Type of Car' is relevant, then the 'buys' relationship is redundant. For example, if a garage wants to keep a list of all customers who have bought a 'Type of Car' from the garage, it can be derived from the set of 'Orders'. However, the garage may wish to do a survey of the local population to see what 'Type of Car' they have bought in the past from wherever. In this case, 'Client' (maybe a subtype of 'Potential Client') 'Buys Type of Car' can exist independently of 'Order'.

12.2.3   Time

Deciding on the view of time for a model is extremely difficult. For example, a snapshot view can be: an 'Hotel Room' (at any one time) contains none or one and only one 'Guest' (let's assume that it is a single room). This is shown in Figure 12.2.

Figure 12.2
A snapshot view of
an hotel room.

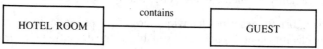

Figure 12.3
A long-term view of
an hotel room.

However, the long-term view is an 'Hotel Room' (over a period) contains one or more 'Guests' (but never at the same time). See Figure 12.3. This view of time may have to change between the high-level model and the later models to be used for design. A good example is when one initially wishes to explain the business but then wishes to ensure a historical record. For example, the business view may be that:

> A 'Customer' travels on one 'Bus' (at any one time)
> A 'Bus' contains many 'Customers' (assume the golden age of transport)

This is shown in Figure 12.4.

Figure 12.4
A view of buses and
customers.

However, we may wish, in the final design, to hold information relating to all customers on all bus trips:

> A 'Customer' travels on one or more 'Buses' (over a period of time)

This is shown in Figure 12.5.

Figure 12.5
A historical view of a
bus travel.

NOTE: In this historical many-to-many case, it is quite likely that we would want to hold additional information about the specific 'Customer travels on specific Bus', e.g. fare paid, comments on service, number of bags carried. Hence, the many-to-many relationship would have to be resolved with an associative entity (explained later).

More about time

In the detailed Entity Model, it is extremely important to consider the relationships at specific times — such as at start up. For example, when a bus sets off it has no passengers and it may be necessary to model this. Another, better, example is when considering an Entity Type which can have existence without the related Entity Type(s), e.g. a house and its relationship with occupants.

When the house is being built, before it is sold, and perhaps at other times, it is unlikely to have occupants, and so it is important to model this by declaring that:

> Each 'House' has zero, one, or more 'Occupants'

Figure 12.6
Houses and
occupants.

This is shown in Figure 12.6. NOTE: This may also be defined as 'Each House may have one or more Occupants' — the *may* explicitly stating the case of 0. I think that 'may' sounds nicer than zero.

However, in this example, it is ambiguous as to whether this is a snapshot in time or information to be held over a period. A snapshot models the occupants as being the people currently living in the house (e.g. to decide whether or not they are to pay the Community Charge). A historical view models the occupants over a period of time (e.g. the house's owner keeping a list of tenants). This leads to the conclusion that information relating to time must be documented, and ambiguities cleared up.

Yet more about time

Time can be modelled explicitly if this helps the understanding. An example is tax discs on a fleet of company cars:

Each 'Month' zero, one, or more 'Car Tax Discs' require renewal

This is shown in Figure 12.7.

Figure 12.7
Explicit modelling of
time.

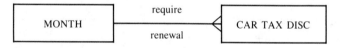

Another aspect of time is that attribute values could change over time. An example is the documents relating to a piece of work for which a proposal has been written. In general, the statement of work in the proposal is modified in order to produce the agreed statement of work for the contracted work. If the original statements are not of interest, then one could have one, and only one, Entity Type 'Statement of Work'. However, if we must keep the original statement as well as the last, then we need to model a one-to-one relationship:

Each 'Proposed Statement of Work' results in zero or one 'Agreed Statement of Work'

This leads to the interesting area of versioning. It may be necessary to hold the interim statements of work such that:

Each 'Proposed Statement of Work' has zero, one, or more 'Interim Statement of Work'

One or more of the attributes of 'Interim Statement of Work' must explicitly or implicitly state whether or not this 'Interim Statement of Work' is the final version.

Being the final version may be important enough to warrant some additional attributes such as 'date signed off', thus creating a subtype of 'Interim Statement of Work', called, say, 'Final Statement of Work'.

12.2.4 *Historical records*

Attributes on the Entity Type can be used to log data at events whenever the Entity Type can only pass through a fixed number of events. Take, for example, a 'Works-authorisation' document where we want to log the workers who reject the work before someone accepts it. We are told that after the third refusal the work package is to be dropped (or revised to a new Entity Type instance). We could define three attributes, namely: 'Worker1-signature', 'Worker2-signature' and 'Worker3-signature'.

A more complicated example is where we wish to log all changes made to an Entity Type instance, including who created it, and we have very little idea on how many changes will be made.

Example: Historical records 1

Take a person's credit-rating record which contains name, address, current debts and credit rating. It may be necessary to keep a record of all the changes made to this record and by whom. As the volume of changes is unknown, it is impossible to store the relevant information alongside the 'latest' data, and so a new Entity Type is required, say, 'Change-record'. Hence, for each 'Credit-rating-record' there may be one or more 'Change-records'.

Now, what should be stored with the 'Change-records'? Certainly, there will be some administrative information such as: 'id of person making the change', 'date and time of the change', 'process performing the change'. Then we need a record of the change.

Here we have to make some decisions. If we want to log any change to any part of the record, the 'Change-record' is logically just a copy of the 'Credit-rating-record' with a further 'administrative' attribute, namely 'pointer-to-fields-changed'. However, if the business knows which fields it wants to log and these fields are a significantly small subset of all the fields (or some other criterion), then at the analysis stage the Entity Type 'Change-record' could be subtyped with a classifying attribute of 'Type of Change', with the subtypes holding the relevant changed attributes.

Then an Entity Life History (ELH) can be determined, mapping low-level activities (at the analysis stage) against the 'change-field' attributes of the 'Change-record' (i.e. all the attributes which match attributes in the original record).

So we have a change/read/update/delete (CRUD) matrix (or we may put in actual values):

	attribute-1	attribute-2	attribute-3
activity-1	C		
activity-3	R		
activity-29		R	
activity-10			U
activity-50	D		

Example: Historical records 2

If the business knows exactly which data items it wants to keep a historical record about (and nothing else), then the implementation, and sometimes the modelling, becomes easier.

Let's take mortgage repayments. A mortgage record belongs to a house/owner partnership. Certain things are important: mortgage number, original loan details, outstanding amount at date and current monthly repayment figure. The lender will also wish to keep track of the payments expected in any one month, and the payment received.

A way of describing this may be: for each 'Mortgage-detail' there will be one or more 'Payment-records'. 'Payment-record' has the following attributes: 'month/year of record', 'payment expected', 'payment received' and 'date payment received'.

With a complete set of 'Payment-records', and the original 'Mortgage-detail' record, there is every possibility that the complete history of the mortgage can be ascertained. This is easy because there is a well-defined set of attributes changing every time.

Example: Historical records 3

On a computer system, the DP manager wants to keep a history of when the memory size, the exchangeable discs and the power supply are updated. The computer system 'entity type' is quite 'long', as it lists all the peripherals on the computer (some 230).

At the highest level (Business Analysis) we, perhaps, would describe an Entity Type 'Computer-system', with attributes holding information such as: current

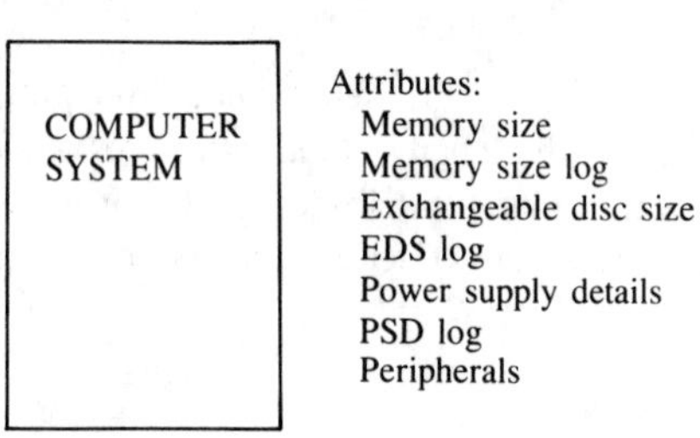

Figure 12.8
View at the Business
Analysis Phase.

COMPUTER
SYSTEM

Attributes:
  Memory size
  Memory size log
  Exchangeable disc size
  EDS log
  Power supply details
  PSD log
  Peripherals

memory size, memory size log, exchangeable disc size, exchangeable disc size log (eds log), power supply details, power supply details log (psd log) — see Figure 12.8.

During the analysis phase we would want to expand this, making the 'log' attributes into relationships to one or more specific log records, as shown in Figure 12.9.

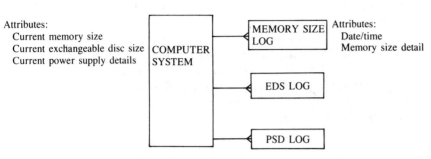

Attributes:
  Current memory size
  Current exchangeable disc size
  Current power supply details

Attributes:
  Date/time
  Memory size detail

Figure 12.9
View at the Activity and Data Analysis Phase.

Figure 12.9 reads as follows:

> Each 'Computer System' has a log of one or more 'Memory-size-log'
> Each 'Computer System' has a log of one or more 'Exchangeable-disc-size-log'
> Each 'Computer System' has a log of one or more 'Power-supply-details-log'

Each new Entity Type will have a set of administrative details (e.g. date/time) and the updated data item information. This can then be 'simplified' into subtypes, as shown in Figure 12.10.

Figure 12.10 is read as follows:

> Each 'Computer System' may have one or more 'Modification-records'
> The 'Modification-record' has three subtypes:
>   Memory
>   Exchangeable
>   Power-supply

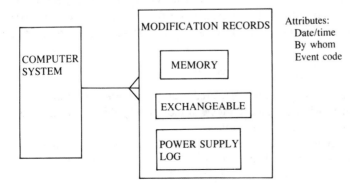

Attributes:
  Date/time
  By whom
  Event code

Figure 12.10
Simplifying Figure 12.9 by using subtypes.

The 'Modification-record' has a set of administrative attributes (date/time, by-whom) plus a classifying attribute (e.g. event code) in order to determine the relevant subtype information.

Development and implementation

During the Analysis and Design phases we are concerned with showing what data must be stored, not how they will be stored. Even during the design, the development details on a particular hardware configuration need not be addressed (but could be if the final configuration is known, agreed and will *never* be changed).

During the implementation phase, the developer makes decisions based on the sizing information given in the Analysis and Design phases. In the case of the 'Modification-record', with subtypes given in example 3 above, the developer may make the 'Modification-record' either:

1. One 'file';
2. *n* 'files', where *n* is the number of subtypes; or
3. *m* 'files', where *m* is less than *n* and some of the subtypes are merged together, most probably for some physical speed/size consideration.

How does the developer decide how to implement the given Entity Type? He or she will have two additional pieces of information:

1. The activities versus attributes 'matrix'.
2. The physical database being used.

The developer will use his/her expertise to decide which is more critical: the speed at update, the speed of on-line access, the speed of the production of reports, or the best use of disc space. Once this decision has been made, the physical database design is chosen which meets the most critical need.

For example, all the subtypes associated with the amendment records could be stored in one file if the smallest physical record that can be created on the database is larger than the required data storage. If space is not a problem, and everything works quite quickly, then it may be easier to get rid of 'small' amendment records and simply copy out the whole, original Entity Type and place a special code in unchanged fields and the changed data in changed fields. This can have enormous benefit in modularising the final program suite, as the structure of the data is defined only once — but, of course, it is wasteful of space.

Multiple relationships

The examples so far have considered history from only one point of view. More often, history is associated with one or more Entity Types, and the history search can be initiated from either end.

For example, a piece of cargo is booked on a flight, and so the flight has had the piece of cargo booked on it. We may need to discover flights that the piece of cargo was booked on over a period of time and/or what pieces of cargo (or

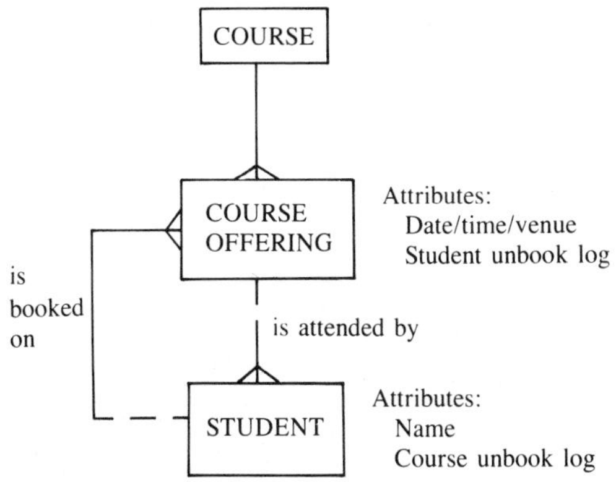

Figure 12.11
Course offerings and
students.

at least certain information such as weight) were booked on a certain flight but
were never flown for some reason.

In the Business Analysis phase of the case study, the Entity Type 'Course
Offering' has been defined as being an instance of running a 'Course'. 'Students'
may or may not attend 'Course Offerings' (see Figure 12.11).

> Each 'Course' will have one or more 'Course-offerings'
> Each 'Course-offering' may be attended by (has latest booking by) one
>   or more 'Students'
> Each 'Student' may be booked on one or more 'Course-offerings'

Important, but ill-defined attributes of 'Course Offering' and 'Student' are 'student
unbooked log' and 'course unbook log' respectively.

During the Analysis Phase we discover the need for the following reports:

1. What courses have had more than five unbookings (cancellations)?
2. How many times does a student book on a course but then unbooks?
3. Report the staff members who fail to rebook students on any other course
   at the time of unbooking.

The data required to support these information needs would appear to include:

1. Name of staff member who performs the booking/unbooking.
2. Data about any new booking made at the time of the unbooking.

After careful analysis the data model is amended, adding in a new Entity Type
'Course-offering-amendment-record' (C-O-A-R-), as shown in Figure 12.12.

> Each 'Course-offering' has one or more 'C-O-A-R-s'
> Each 'Student' may be associated with one or more 'C-O-A-R-s'
> Each 'Course-offering' may be attended by (latest booking by) one or
>   more 'Students'

The Entity Type 'C-O-A-R-' has two subtypes: 'Book' and 'Unbook'. The
'Unbook' subtype has two important attributes, namely: 'reason for unbooking'
and 'details of any new booking'.

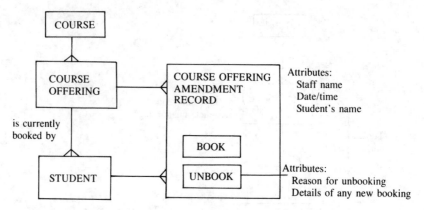

Figure 12.12
Amended Data
Model.

The relationship 'may be attended by (latest booking by)' is potentially redundant, as it can be derived by evaluating the C-O-A-R- records, of which there will always be one between C-O-A-R- and a student on the specific course (the original booking). However, there may be many C-O-A-R-s, and so this may get long-winded. Therefore, showing the direct 'latest' relationship is useful. To get a feel for how the required reports listed above are generated, the initial process logic (i.e. at Analysis Phase) is given below:

Report 1
What courses have had more than five unbookings?

        FOR EACH Course
        FOR EACH Course-offering
            Count the number of 'unbook' C-O-A-R-s
            IF count > 5 THEN print Course, Course Offering
            END IF
        END FOR
        END FOR

Report 2
Which students booked on a course but then, to date, have consistently unbooked themselves and are currently not booked on the course?

        FOR EACH Student
        IF Student has some C-O-A-R-s THEN
            FOR EACH unique Course-offering
                Count 'booked' and 'unbooked' records
                IF unbooked > booked THEN
                Print Student 'booked but didn't go on' Course
                END IF
            END FOR
        END IF
        END FOR

Report 3

Find the staff who most often fails to rebook a student who unbooks:

```
FOR EACH 'Unbooked' record in C-O-A-R-
IF 'New Booking' is not filled in THEN
    Add 1 to count for Staff-name
END IF
END FOR
Find Staff-name with highest count
Print "Fire ", Staff-name, "!"
```

Having done all this, we may then decide that the use of 'unbook' was clumsy. We could decide to rework the deliverables and use something like 'student cancellation log' and 'course cancellation log' instead of 'unbooked' and 'unbookings'. If the deliverables were produced by hand, I doubt if one would have had the energy to do this. However, if the deliverables are on a suitable Case tool, then perhaps one would.

12.2.5 *One-to-one relationships*

One-to-one relationships do not, in general, feel correct, and so each one should be assessed to consider whether the related Entity Types can be merged.

The most likely case for a correct one-to-one relationship is when either of the joined Entity Types can exist on its own without the other. For example, a company may produce a statement of work directly (say for an internal project) or as a result of a proposal. Hence, there is a one-to-one relationship between 'Proposal' and 'Statement of Work', but there may be proposals without a statement of work (lost the business) and there may be a statement of work without a proposal (internal project).

In general, other cases for not merging Entity Types are:

1. Most attributes and relationships differ. For example, an order gives rise to a delivery. Here, 'Order' may have information about the goods required and 'Delivery' will have customer's name, address, time for delivery, truck being used, truck driver's name, etc.
2. Corresponding attributes can take different values. For example, a proposal may result in a contract where certain attributes may have different values, such as price, payment terms, date of starting, etc.

One further case is when management/users prefer to see the two names appear on the model; although it may be sensible to try to win them over to the required consolidated Entity Type. The two Entity Types *should be merged* when:

1. Both Entity Types *must always* exist together. For example: each 'House' has one and only one 'Location' (where 'Location' can be a lot number, address or OS reference). 'Location' is probably an attribute of 'House'.
2. Identifiers are identical. For example, 'Employee' becomes 'Retired

Employee': both Entity Types have the same id (e.g. name or personnel number).

3. Entities have many identical attributes and relationships. For example, 'Aircraft' may be a 'Truck', where 'Aircraft' has attributes: number of passengers and freight configuration, and 'Truck' has attributes: freight configuration. Here, 'Aircraft' and 'Truck' should be combined into one Entity Type such as 'Freight Carrier' if the model is freight-orientated, or perhaps the definition of 'Aircraft' extended to include 'Truck'.

4. Main difference is time. For example, in the 'Proposal results in Contract' example, it may be that we wish neither to keep a historical record, nor to have differing use of attributes — all that is required is the latest status of the statement of work and conditions. In this case, the two Entity Types could be merged into one called, say, 'Potential Contract' with an attribute such as 'Status' (proposal, refining, contract).

12.2.6   *Many-to-many relationships between different Entity Types*

Many-to-many relationships tend to be confusing — leading to ambiguous meaning. As the main idea of modelling is to make things clearer, it follows that many-to-many relationships should be resolved whenever possible.

Within the high-level Data Model, it is best to show only the main relationships, together with the most common cardinality. This would tend to mean showing one-to-one and one-to-many relationships only (that is, forget the many-to-manys and the zero cases). The only exception is when the many-to-manys (and maybe the zeros) help make the analysis clearer from a senior manager's/user's point of view.

In many cases, the many-to-many relationships mean completely different things in each direction. For example, a Data Model diagram which shows a vehicle with a many-to-many relationship to a person leaves unclear what the relationships are. A good sign of this confusion can usually be spotted when trying to name the relationship. The more wishy-washy the name, the more likely that there is some uncertainty of the relationships involved.

Such a many-to-many relationship can usually be resolved by taking each of the Entity Types in turn, placing the word 'each' in front of them, and then looking for the one-to-many relationship to the other Entity Type. Therefore, in the above example:

> Each 'Vehicle' may be driven by one or more 'Persons'
> Each 'Person' may own one or more 'Vehicles'

(One would have to make it clear whether this was a snapshot view or a view over time.) The diagram would be redrawn, replacing the one many-to-many relationship with two one-to-many relationships, each named succinctly in the one-to-many direction.

In a truly many-to-many relationship, such as the one above, the many-to-many, once resolved as shown above, may be correct. However, we may wish to hold additional information about the relationship. For example, we may wish to hold the miles driven in each vehicle by each of the possible drivers. For example, let us state that the many-to-many relationship contains the following information:

Vehicle	ABC123A	is driven by	Leslie and Roger
	ABC124A	is driven by	Steve
	ABC125A	is driven by	Steve and Theo
	ABC126A	is driven by	Roger

This is shown in Figure 12.13.

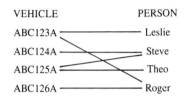

Figure 12.13
A many-to-many relationship.

We now insert a new Entity Type to hold 'Miles Driven', such that:

Each 'Vehicle' is driven one or more 'Miles Driven'
Each 'Driver' drives one or more 'Miles Driven'

which leads to:

Each 'Driver' has driven 'Miles Driven' in 'Vehicle'

So the table, taken from a driver's viewpoint, would be:

Leslie	has driven	04 miles	in ABC123A
Steve	has driven	17 miles	in ABC124A
Steve	has driven	29 miles	in ABC125A
Theo	has driven	100 miles	in ABC125A
Roger	has driven	123 miles	in ABC126A
Roger	has driven	12 miles	in ABC123A

This is shown in Figure 12.14.

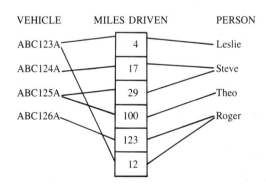

Figure 12.14
Table of miles driven.

For the sake of completeness, this new Entity Type ('Miles Driven') is quite often called an *associative* Entity Type, or an *intersection* Entity Type.

12.2.7 Many-to-many involuted relationships

A many-to-many involuted relationship is when an Entity Type has a many-to-many relationship to itself. I find these particularly alarming on a diagram. As an example, 'Part' is made from one or more 'Parts', and 'Part' is used in one or more 'Parts'.

If this is all the information that is needed, then, just as in the many-to-many relationship between different Entity Types case, all is well. However, one must check whether additional information needs to be stored. In the above example, it may be necessary to store the number of parts involved and/or the order of the parts. Hence, there may be a need to generate an *associative* Entity Type.

In summary: (1) involuted many-to-many relationships are extremely confusing to the senior manager/user, and (2) they should be resolved in the same manner as many-to-many relationships between different Entity Types.

12.2.8 Integrity conditions

When the completed program (processes) use and search the implemented database, they will tend to depend on certain aspects of the database structure — as defined by the final (lowest) Data Model. The fact that data are stored according to the defined data structures is termed 'database integrity'. Some common integrity conditions are:

1. If the relationship between two Entity Types is mandatory, then the process will expect to 'traverse' at least one such link.
2. If the relationship is optional, then the process will need to know when the link exists, allowing it to 'traverse' the link.

In the optional relationship case, the existence of the link may depend on one or more attribute values and/or on the existence of one or more other links. For example:

> A 'Vehicle' may be driven by an 'HGV-driver'

> An HGV-driver participates in the pairing if the attribute 'type' on the Entity Type 'Vehicle' has the value of 'heavy goods'.

As a second example:

> Each 'Pupil' may be registered taking an 'Exam'
> Each 'Pupil' may have a set of 'Examination Papers'

> Then a pupil will have a set of examination papers if the pupil is registered taking the exam.

There may be other integrity conditions that matter, especially relating to data distribution. It is important that all of these are made clear before design starts.

12.3   Objectives of the Design Phase

At the end of the Data and Activity Phase, one has:

1. A *Data Model* comprising:
    *   *Entity Types:* name, synonyms, definition, occurrences, growth, attributes, identifiers, exclusive relationships
    *   *Relationships:* name, definition, integrity conditions, optionality %, cardinality minimum/maximum/average
    *   *Entity Subtypes:* as Entity Types plus classifying attribute value
    *   *Attributes:* name, synonyms, definition, integrity conditions, optionality, domain, length, permitted values or range, default value
2. A set of *activities* on an *Activity Hierarchy Diagram* and a number of *Data Flow Diagrams*. Each activity at the lowest level (that is, an activity that cannot be further decomposed) comprises:
    Name
    Definition (free text description)
    Structured description (especially for activities that are not well understood either procedurally and/or their effect on the data) — optional
    Frequency
    Growth
    Proposed mechanism
    Expected effects
    Inputs
    Outputs

The objectives of the Design Phase are:

1. To reach agreement with the users on the ways in which the users will interact with the system, and determine a set of preliminary/outline screen designs. From these, a set of detailed screen designs will be produced.
2. To design a 'business' system which will support a selected set of activities (as defined and selected during the Analysis Phase).
3. To complete the design to the lowest level possible without prejudging technical issues.

The tasks required to complete the Design Phase are:

1. Define and describe the procedures that will carry out the lowest-level activities.
2. Complete the Data Access Flows. (These may have been started during Analysis to help produce the structured English descriptions — more on this later.)
3. Produce a Logical Data Structure — which identifies the keys required for the Entity Types in the Data Model.

This is summarised in Figure 12.15.

During the Design Phase, the Data and Process Models from the Analysis Phase may have to be modified in order to rectify errors of thinking; at the end of the Design Phase, they will be shown to be complete.

Figure 12.15
Analysis Phase to
Design Phase
mapping.

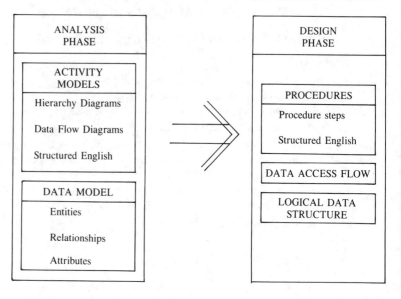

From the Design Phase deliverables, the Implementation Phase will produce the computer code in some implementation language (which may, of course, be remarkably similar to the structured English used in the procedure descriptions, e.g. Macintosh HyperCard language), the screen generation and update code, and the database control code (definitely the generation code, but may also be some form of database manager, e.g. to ensure integrity).

12.4   Recap on the concept of lowest-level activities during Analysis Phase

During the Activity and Data Analysis Phase, the activities have been progressively refined from the highest-level view to the next-level view. No guidance has been given as to where to stop refining the Activity Hierarchy Diagram, only that the lowest-level activities must be of interest to the group of reviewers at any stage.

However, by the end of the Activity and Data Analysis Phase, one must identify the lowest-level activities. A lowest-level activity is an activity which:

1. Has some significant effect on the data identified in the Data Model to date.
2. Must be completed (that is, once started must finish) in order to leave the data within the Data Model in a consistent state (in a business sense).
3. (Similar to (2)) when completed will leave the data in a consistent state, and so does not need further actions in order to do this.

During the Activity and Data Analysis Phase, each lowest-level activity will be detailed using some form of structured English. The style will be:

1. Each lowest-level activity is detailed in one, and only one, piece of structured English.

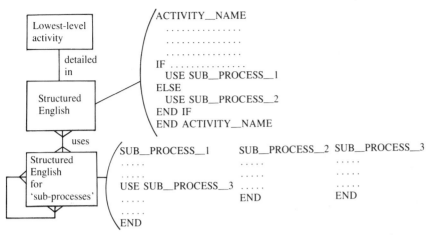

Figure 12.16
Review of the
Analysis Stage –
lowest-level activities.

2. For the sake of clarity, the structured English may make reference to sub-processes (like subroutines), with each subprocess itself defined in structured English (see Figure 12.16).

12.5 Mapping lowest-level activities to procedures

Each lowest-level activity defined during the Analysis Phase will now be supported by one or more procedures. Procedures define how the 'what' defined in the process is to be done.

In an ideal world, each lowest-level activity would be transformed into one procedure. In reality, there are a number of possibilities:

1. One activity may be mapped to one procedure.
2. One activity may be mapped to many procedures in order to support alternatives.
3. One activity may be mapped to many procedures, each procedure performed after another (i.e. serially).
4. Two or more activities may be supported by, and so mapped to, one procedure.

Furthermore, additional procedures may be identified in order to support any necessary or enhanced operations not identified at the Analysis Phase.

One activity to many procedures in order to support alternatives

Alternatives to support things like:

Fall-back situations
Different operator types and/or operating environments
Various work situations need supporting

An example covering most of these is: activity 'Take Order' may require several alternative procedures:

'Receive Mail Order'	(work situation 1)
'Take Telephone Order'	(work situation 2)
'Take Counter Order'	(work situation 3, operator type 'own staff')
'Take in-shop Order'	(work situation 3, operator type 'shop staff')
'Correct Order'	(fall-back situation 1)

One activity to many procedures — serial procedures

This is when an activity requires two or more procedures to be performed one after another in the specified order. The reasons for this may be because:

1. One or more of the procedures will occur at a different physical location.
2. One or more of the procedures will be performed by different personnel.
3. There is a time lapse between the procedures.

Many activities supported by one procedure

This occurs when two or more activities are always performed in series, or when a procedure can be generalised in such a way that it can support several activities.

A good example is when the data requirements for the many activities are identical and the user has to choose between a number of one key actions. For example, the activities 'Take Order', 'Amend Order' and 'Cancel Order' could all be supported by the procedure 'Maintain Orders' with a user action menu of '1 for Take Order', '2 for 'Amend' and '3 for Cancel Order', or whatever.

Additional procedures

As one maps activities to procedures, new areas requiring support will be found. Such areas will be necessary for, or enhance, the operation of the final system, and are unlikely to have been found in the Analysis Phase. Some examples are:

> For control menus
> For maintenance of the fixed/pre-defined information
> For correction of data entry errors
> For regular internal reports and enquiry screens
> For on-line testing
> For audits and the like
> For deleting data
> For transition from the current system(s) to the new one

Each procedure may consist of one or more *procedure steps* (see Figure 12.17).

This is purely a simple structured design device which can be useful for screen handling. Wherever possible, it is recommended that one screen is supported totally by one procedure step. In the days of windows and multitasking, this recommenda-

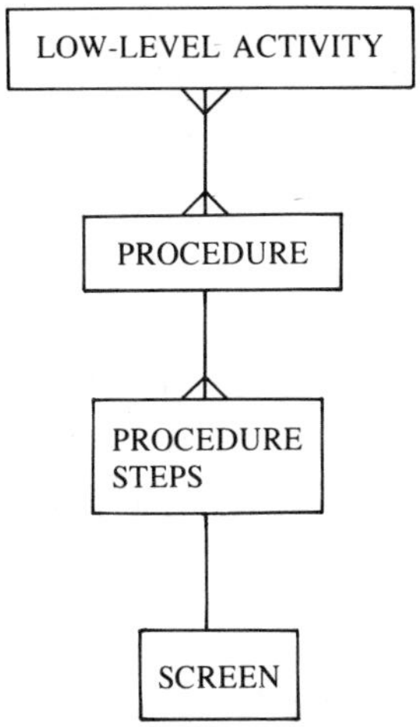

```
┌─────────────────────────┐
│    LOW-LEVEL ACTIVITY   │
└─────────────────────────┘
            △
            │
      ┌───────────┐
      │ PROCEDURE │
      └───────────┘
            △
            │
    ┌───────────────┐
    │   PROCEDURE   │
    │    STEPS      │
    └───────────────┘
            │
       ┌──────────┐
       │  SCREEN  │
       └──────────┘
```

Figure 12.17
Process and
procedures.

tion may need further clarification. However, I do not know of any well-defined techniques for providing this clarification. Suffice to say that the overall motto must be 'Keep it simple'.

12.6   Steps in defining procedures

Step 1   Map lowest-level activities to procedures.
Step 2   Identify any obvious additional procedures (mainly for control menus).

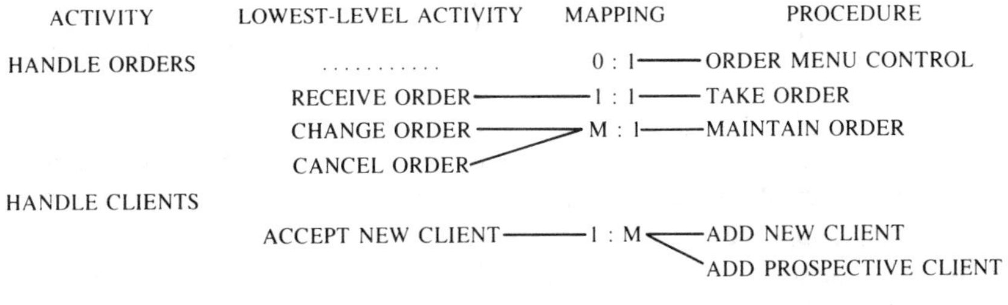

```
ACTIVITY          LOWEST-LEVEL ACTIVITY   MAPPING        PROCEDURE

HANDLE ORDERS        . . . . . . . . . .    0 : 1————ORDER MENU CONTROL
                  RECEIVE ORDER————1 : 1————TAKE ORDER
                  CHANGE ORDER————M : 1————MAINTAIN ORDER
                  CANCEL ORDER

HANDLE CLIENTS
                  ACCEPT NEW CLIENT————1 : M————ADD NEW CLIENT
                                                ADD PROSPECTIVE CLIENT
```

Figure 12.18
Mapping activities to
procedures, and
identifying additional
procedures.

Step 3   Consider the outline logic and user interface to the identified procedures.
Step 4   Break down each procedure as necessary into procedure steps such that either:
- Each procedure has no procedure step and only zero or one screen associated with it, or
- Each procedure has one or more procedure steps associated with it, and each procedure step has zero or one screen associated with it.

NOTE: Step 4 is not absolutely necessary — but it does help to modularise the design and so clarify thinking.

An example of a breakdown is shown in Figure 12.19.

PROCEDURE	PROCEDURE STEP	SCREEN
ORDER MENU CONTROL	None	SCREEN__ORDER__MENU
TAKE ORDER	Initial__questions	SCREEN__TAKE__ORDER__1
	Supplementary__queries	SCREEN__TAKE__ORDER__2
	Order__details	SCREEN__TAKE__ORDER__3
MAINTAIN ORDER	None	SCREEN__MAINTAIN__ORDER
ADD NEW CLIENT	None	SCREEN__ADD__NEW__CLIENT
ADD PROSPECTIVE CLIENT	None	SCREEN__ADD__PROSPECTIVE__CLIENT
none	CORRECT__ORDER	SCREEN__LOCAL__CORRECTION__ORDER
none	CORRECT__CLIENT	SCREEN__LOCAL__CORRECTION__CLIENT

Figure 12.19
Procedures/procedure
steps/screens.

Step 5   Define each procedure/procedure step using one of the following:
   • Structured English refined from the process definitions;
   • A new structured English — a little nearer to the implementation style (i.e. pseudocode);
   • An implementable computer language such as 4GL or a database language such as is available with DataEase.

Step 6   Identify additional procedures (e.g. for correction of data entries).

12.7   Overview of the steps in the Design Phase

Step 1   Take a copy of the Analysis Phase Data Model (or at least that part of the Data Model that is relevant to this part of the system).

Step 2   For each lowest-level activity, produce a Data Access Flow on top of the Data Model. This may be a refinement of work already carried out during the Analysis Phase. The Data Access Flow shows the entry points into the data and the access volumes (the technique is explained in the next chapter).

Step 3   For each lowest-level activity, produce a Logical Data Structure Diagram based on the Data Model. This shows the required foreign keys (the technique is explained in the next chapter).

Step 4   For each lowest-level activity, make a decision on how they will work: that is, define the procedures needed. How this is done depends very much on the type of system being implemented. However, all systems require information to flow either into and out of the process as a whole, or into and out of the procedures to some external device (such as the screen). Analysing this 'dialogue' flow will enable logic to be grouped into 'input processing output' procedure steps — each requiring some interaction with the 'outside world'.

Step 5   For each procedure/procedure step requiring interaction with the user via the screen, there is a need to define the contents and layout of a specific screen.

Step 6   Identify the additional procedures required now that the design is more detailed — for example, for menu control.

Step 7   Procedures and procedure steps are now defined using a logic description language such as structured English.

Step 8   The Analysis Phase deliverables are reviewed for completeness based on the refined thinking.

12.8   Some basic relational database concepts

Relational databases are made up of *tables*. A table is simply a rectangular set of values; along the top is a set of *field* names labelling the columns. By convention, the first (leftmost) field (although it may be more than one field) uniquely defines a *record* (line/row) in the table. The same field name may not be repeated in any one table, but may be used in more than one table.

The number of fields is fixed at database generation time; however, the number of records is not (logically at least).

Once the *unique key* combination (one or more fields) has been defined, all of the fields (including those in the unique key combination) may be considered as further *keys*, *keys for linkage* (foreign key) or just names for searching/updating purposes.

An ideal mapping of the Data Model to the database structure is:

Entity Type	→	Table
Relationship	→	Linkage key field
Identifier attribute	→	Key field
Attribute	→	Field

12.9   Preliminary data structure design

The preliminary data structure design is the first attempt at designing the database. Basically, the required tables, fields and linkages are defined.

During the Design Phase, one is producing a logical design — one that the users can understand. It is not a physical design: that is, it does not show the physical file organisation nor the performance options.

How tables are actually held, e.g. in separate computer files or whatever, depends very much on the implementation details, and so should be left until the Implementation Phase. Figure 12.20 summarises the processes taking place post design and during implementation.

12.10   Designing the system controls

The work performed during the Analysis Phase and at the start of the Design Phase determines the procedures required to directly support the business needs. There are many other areas which need addressing to ensure that the system will work correctly in the final environment. The main areas to consider are:

Figure 12.20
Data Model/activity
interaction.

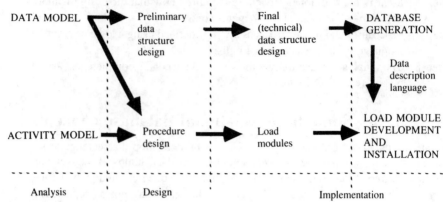

Security
Data integrity
Privacy/accessibility
System availability
Business operational efficiency

Such consideration may result in the need for:

Additional procedures
Additional fields
Alterations to procedures already defined

Some examples are:

Data record creation date
Who last updated the record
On-line deletion not allowed — record flagged
Data consistency checker
Security violation checker

Part 5

Design and implementation:
case study

13 Case study: Design Phase for 'Schedule Trainer for Course'

13.1 Plan

The Project Meeting agreed to continue with designing the area 'Schedule Trainer for Course', and the plan is shown in Figure 13.1.

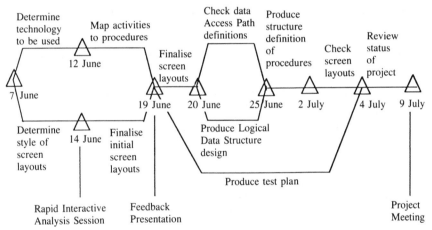

It was decided to have a Rapid Interaction Analysis session to discuss and agree the basic screen layouts to be used, and this has been scheduled for 14 June. A clearer picture of the screens and procedures will be presented at a Feedback Presentation to be held on 19 June. This meeting will be mainly for Implementers, but the users will be represented by one of their keyboard operators. The final design will be reviewed at the Project Meeting to be held on 9 July.

13.2 Technology to be used

RCT has just replaced their ageing IBM PS2s with Apple Macintosh SE30s and 2CXs for hands-on training. It was decided to use similar computers for the proposed support system.

The technology will be monochrome M68030-based Apple Macintoshes, using

Oracle as the database (bought already for training), and Apple HyperCard with HyperTalk scripts to control the user interface and process logic, and Oracle HyperSQL to interface with the database.

It is considered that the HyperCard script language is easy enough to follow at the level used in this set of programs. Within this script language, there is a special command 'execsql' which allows Oracle SQL statements to be obeyed. HyperCard has quite a small line length limit (in order that a whole card may appear on a small screen); hence, most SQL statements have to be split across more than one line. The technique for doing this is to split the SQL statements into phrases, put each phrase on a new line between double quotes and each line (except the last) with the characters '&& ¬ ', where ¬ is the key strokes 'option — return'.

An extension to the standard SQL is the ability to use HyperCard variables in the SQL statement, and these are shown with colons around their name, e.g. ':Mseqno:'. Hence the SQL statement:

> UPDATE table1 SET info1 = 'a variable named required info' WHERE courseid = 'a variable named Mcourseid' AND seqno = 'a variable named Mseqno'

would appear in the HyperCard script as:

> execsql "update table1 set info1 = :required info:" && ¬
> "where courseid = :Mcourseid: and seqno = :Mseqno:"

where required info, Mcourseid and Mseqno are variables already set up with values in the HyperCard script that surrounds the SQL statement.

13.3   Style of screen layouts

After some simple prototypes, the following was decided. Screens would be based around the HyperCard system and that there would be two different screen (card)

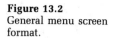

Figure 13.2
General menu screen format.

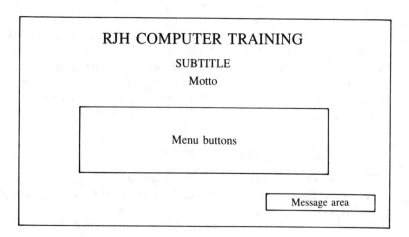

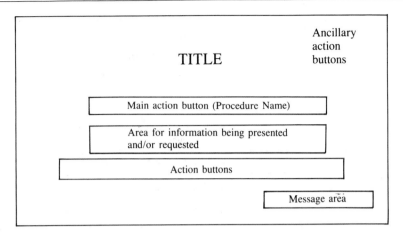

Figure 13.3
General process
screen format.

types: one for menu screens, the other for process screens. The general format
for menu screens will be as shown in Figure 13.2.

The general format for process screens will be as shown in Figure 13.3.

Both screens have two additional buttons, a 'home' button to allow the user
to finish a session, and a 'go back one' button (in the shape of a left arrow) to
allow the user to complete the current process.

13.4   Initial screen layouts

Potential screen 1

Figure 13.4 shows the agreed format for the opening menu.

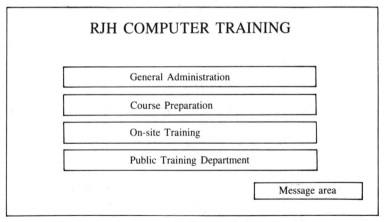

Figure 13.4
Potential screen 1.

Potential screen 14

Figure 13.5 shows the agreed format for the opening menu for the Public Training
Department.

Figure 13.5
Potential screen 14.

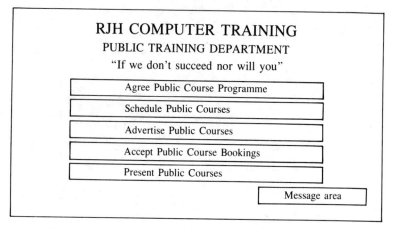

Potential screen 144

Figure 13.6 shows the agreed format for the menu for 'Schedule Public Course'.

Figure 13.6
Potential screen 144.

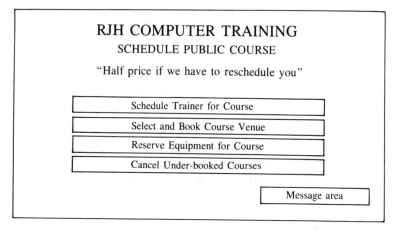

Potential screen 1441

Figure 13.7 shows the agreed format for the menu 'Public Training Department'.

13.5   Map activity to procedures

The activity under consideration is 'Schedule Trainer for Course'. Clearly, the key procedure is 'Schedule Trainer for Unbooked Course'; however, it has been decided that additional 'searches' would be useful, namely:

1.   Look at trainer booking information.
2.   Look at course information.

So the activity will comprise three procedures, as shown in Figure 13.8. The

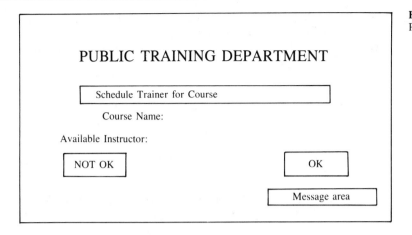

Figure 13.7
Potential screen 1441.

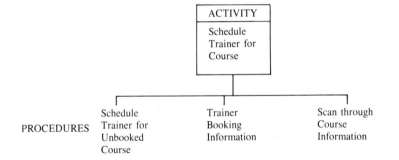

Figure 13.8
Procedures which
make up activity
'Schedule Trainer for
Course'.

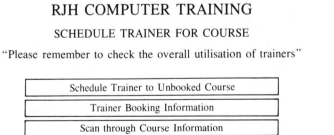

Figure 13.9
Menu screen for
procedures within
'Schedule Trainer for
Course'.

initial version will have stubs for the last two procedures, with a message "Sorry — this has not yet been implemented".

Looking back at the specification phase, we do not have a suitable screen to initiate these procedures:

1. Potential screen 144 is OK and so will be used as is
2. Potential screen 1441 now becomes design screen 14411
3. A new screen, design screen 1441, is to be defined as a menu for the new procedures

Design screen 1441 is shown in Figure 13.9.

13.6   Data Access Path definition for 'Schedule Trainer for Course'

13.6.1   *Some notes on the Data Access Path technique*

The designer uses the Data Access Path technique to ensure that:

1. The proposed activity can access all the data that it needs.
2. The designer can determine the required logic.

In this case study, I have only shown the technique being used in the Design Phase; however, analysts may use it to help clarify 'difficult' activities during the Analysis Phase. The idea is to take the Data Model and work out how the activity will access the first item of data, and then how the activity will gather the rest of the data it needs by traversing the relationships on the Data Model. At the same time, designers can work out how many database records are likely to be accessed. This can help warn the implementers about potential bottlenecks in the final program.

Performed earlier in the life-cycle, the analyst may be able to help the designer/implementer choose the correct technologies for the job: for example, a dedicated hardware database server instead of a mainframe database.

Figure 13.10
Data Access Path definition for 'Schedule Trainer for Course'.

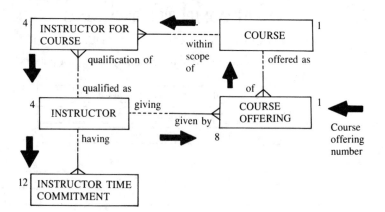

The use of the technique is shown by example for the activity 'schedule trainer for course'.

13.6.2   The Data Access Path definition for 'Schedule Trainer for Course'

The final Data Access Path definition is shown in Figure 13.10. This shows that the data is accessed via 'Course Offering Number' and matches the structured English in Chapter 11, section 11.17 — 'READ COURSE OFFERING USING Course Offering Number'.

One record will be read, hence the 1 beside the Entity Type 'Course Offering'. For each course offering there will be one course which can be accessed via the relationship 'of'. The structured English is:

READ COURSE USING OF COURSE OFFERING

The diagram shows this link with an arrow moving from 'Course Offering' to 'Course'. Then, from 'Course' the activity can read one or more 'Instructor for Course'. On average, there are four 'Instructors for Course' per course, hence the 4 by the box. For each 'Instructor for Course' there is one instructor, and so a total of four is likely to be read.

To find out whether an instructor is free, we have to check whether the instructor is already doing a course in the particular time period. Hence, we move down the relationship 'given by'. On average, we find that each instructor is likely to be assigned to two course offerings, hence the 8 by course offering.

We also have to access the 'Instructor Time Commitments' and, on average, there will be three instructor time commitments per instructor, giving a possible number of accesses of 12.

Hence, for one course offering number there are likely to be $1+1+4+4+8+12$ database accesses, that is, 30 accesses. Any bottleneck is likely to be in reading the instructor time commitments.

As the above has most probably made clear, Data Access Path definitions are also useful in helping one to write the structured English — as opposed to the other way round where, in this case study, we wrote the structured English first. However, it is unlikely that the analyst and/or designer will have enough time or stamina to use this technique on all of the activities. Hence, only those activities which are difficult to understand and/or those which are likely to make heavy use of areas of the database are analysed in this way.

13.7   Logical Data Structure Diagram for 'Schedule Public Courses'

13.7.1   Definition of the technique

Although the Analysis Phase has carefully drawn a Data Model with Entity Types and relationships, real-life databases cannot directly implement relationships.

Pointers from one Entity Type to another Entity Type have to be held in data items associated with the Entity Type. These pointers are quite often called *foreign keys*, and are a copy of the identifier in the Entity Type being pointed to.

The Logical Data Structure Diagram identifies the necessary foreign keys. There are a number of different ways of showing the foreign keys. In this case study, the keys are shown on Entity Types at the 'many' end of a relationship. The technique is demonstrated below for 'Schedule Public Courses'.

13.7.2   The Logical Data Structure Diagram for 'Schedule Public Courses'

The Logical Data Structure Diagram for 'Schedule Public Courses' is shown in Figure 13.11. This diagram is based on the final Data Model generated during the Analysis Phase and shown in Figure 11.3 in Chapter 11.

The Entity Type 'Course Offering' needs keys to 'Course', 'Instructor' and 'Location'. The identifiers for these Entity Types are given in the attribute list generated during the Analysis Phase. This list is given in Chapter 11, section 11.10.

Hence, the required foreign keys to be associated with 'Course Offering' are:

Figure 13.11
Logical Data Structure
Diagram for
'Schedule Public
Courses'.

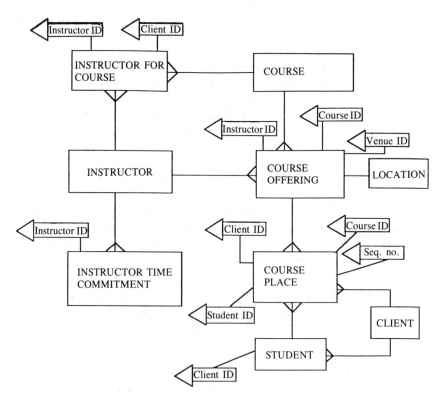

For Course	Course id
For Instructor	Instructor id
For Location	Venue id

These are shown in Figure 13.11. Similarly, 'Instructor for Course' requires foreign keys as follows:

For Course	Course id
For Instructor	Instructor id
and so on	

Foreign keys for 'Instructor Time Commitment' are:

For Instructor	Instructor id

Foreign keys for 'Course Place' are:

For Course Offering	Course id + Sequence number (Seq. no.)
For Client	Client id
For Student	Student id

Foreign keys for 'Student' are:

For Client	Client id

13.8 Structured English definition for procedure 'Schedule Trainer for Course'

*/Find a course offering that requires a trainer (or instructor)
USE Course Offering
LOCATE FOR Course Offering.instructid = ' '
IF no such record THEN END PROCESS
STORE Course Offering.courseid TO Mcourseid
STORE Course Offering.seqno TO Mseqno
STORE Course Offering.weekno TO Mweekno
*/Find number of weeks required to book it
USE Course
LOCATE FOR Course.courseid = Mcourseid
IF no such record THEN ERROR
STORE Course.noofwks TO Mnoofwks
*/Has this course got any qualified instructors?
USE Instructor for Course
FOR EACH Instructor for Course WHERE Instructor for
Course.Courseid = Mcourseid
    IF no such record THEN PRINT "No qualified instructors yet";
    END PROCESS
    STORE Instructor for Course.Instructid TO Minsid

```
*/Get instructor details
USE Instructor
READ Instructor USING Qualification_of
*/Has this instructor got any free time for this course?
USE Instructor Time Commitment
LOCATE FOR Instructor Time Commitment.weekno = Mweekno
IF no such record THEN reject this instructor; LOOP FOR
Repeat the Locate and the IF for Mweekno +1, +2, etc depending
on Mnoofwks
*/So if here we have found a suitable instructor
CONNECT Instructor TO Mseqno USING giving
FOR EACH week of the Course Offering
CREATE Instructor Time Commitment USING Instructor having
END FOR
END PROCESS
END FOR
PRINT "No free instructors for this course"
END PROCESS
```

13.9   Finalise screen designs

This would normally be a list of the screen designs. However, in this case, there are no further major changes to the design considered during the specification phase.

13.10   Produce test plan

13.10.1   Dialogue flow

The dialogue flow in Figure 13.12 shows how the test system moves from log-on through test to the area being tested, namely 'Schedule Trainer for Course' and specifically 'Schedule Trainer for Unbooked Course'.

Figure 13.12
Dialogue flow for
'Schedule Trainer for
Course'.

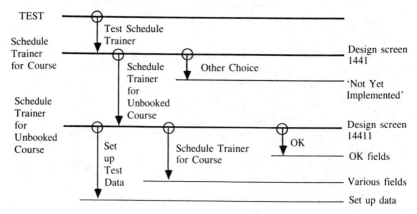

The dialogue flow technique has not been addressed before in this book. It is a simple technique for showing how users navigate between procedures and screens.

The following is a narrative to Figure 13.12. The area of interest is activated from the activity 'Test'. From this initiation, there is at least one choice, namely 'Test Schedule Trainer'. Note that this dialogue flow does not show any other choices, which most probably means that this dialogue flow is, or is about to be, a part of a larger Dialogue Flow Diagram.

So, selecting 'Test Schedule Trainer' enters into the area of 'Schedule Trainer for Course' and the user is shown design screen 1441. Two possible choices are of concern to us:

1. Selecting 'Schedule Trainer for Unbooked Course'.
2. Any other choice.

If the user does not choose 'Select Trainer for Unbooked Course', the system displays 'Not Yet Implemented'. It is not clear from the dialogue flow, but we then want the design screen 1441 to reappear. The structural notation of placing a circle at the start of the arrow line means 'return here' when the user finishes the lower-level procedure.

If the user chooses 'Select Trainer for Unbooked Course', we enter that area of the program and design screen 14411 appears. The dialogue flow now has three options available to the user, namely:

1. Select 'Set up Test Data'.
2. Schedule Trainer for Course.
3. OK.

The dialogue flow shows that for each of these choices a set of low-level procedures is initiated — variously named as 'Set up Data', 'Various Fields', 'OK Fields'. We cannot deduce what these procedures are doing; the dialogue flow is just giving us the hierarchy of the procedures and the supporting screens.

13.10.2   Short-cut

It was decided to test 'Schedule Trainer for Unbooked Course' directly — at least to keep the managers happy. Hence, design screen 14411 was modified slightly and expanded to set up the data for the test. The new format is shown in Figure 13.13.

13.10.3   Test plan

To test that the system can correctly identify unscheduled course offerings, find a 'free' trainer for that course offering and then update the database to assign the free trainer to the course offering. To test that the system correctly identifies the situation when there are no course offerings requiring a trainer and when, although there are course offerings requiring a trainer, there are no free trainers.

Figure 13.13
Modified design
screen 14411 (for
testing purposes).

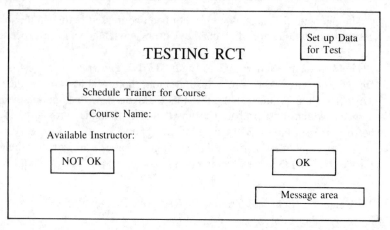

The test data will be set up in a new set of tables and the 'Schedule Trainer
for Unbooked Course' procedure run through manually on the screen.

13.10.4   Test data

The following test data will be required:

1. Set up five trainers who can run all courses.
2. Set up four courses each lasting one week, two weeks, three weeks, and two
    weeks respectively.
3. Set up eight course offerings such that all five trainers are selected one after
    another, then a few extra ones, plus one course that is impossible to schedule.

(This is not a comprehensive set of tests.)

13.11   Review status of project

Having reviewed this test plan and use of screens, it is clear that my testing strategy
is not very structured after all. Before testing the rest of the system, the screen
and procedures for 'Test' needs revising. In particular, this is where the test data
should be set up, not on the final application screen. The screens at the lower
levels *must* look *exactly* as they will in the final complete system.

14 Case study: Implementation Phase for 'Schedule Trainer for Course'

14.1 Plan

Despite the criticisms of the screens (see Chapter 13, section 13.11), it was decided to move ahead into implementation for this one area. The lessons learned will be addressed when the project performs the Data and Activity Analysis for the other to be agreed areas for automation.

The Implementation Phase will take place from 9 July until 23 July, with the program suite ready for final testing and quality evaluation in the period 17–20 July. The plan is shown in Figure 14.1.

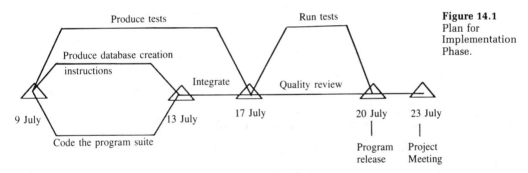

Figure 14.1 Plan for Implementation Phase.

Post-implementation note

Assuming that the Project Meeting on 23 July gives the go-ahead, the program suite will be evaluated (beta tested) until 17 August by a group of users. A special Project Meeting with the managers and users will take place on 31 August, when the state of the project will be reviewed.

14.2 Database creation program

Five tables need creating, and the SQL script for doing this is as follows:

```
CREATE TABLE Course    (Courseid char(10),
                        Coursenm char(40),
```

	Noofwks number(1), Noofdays number(2), Maxplaces number(2), Equipnotes char(70))
CREATE TABLE Instruct	(Instructid char(3), Surname char(15), Firstnm char(8))
CREATE TABLE Offering	(Courseid char(10), Seqno number(3), Weekno number(3), Startday number(1), Status number(1), Geocode number(2), Instructid char(3), Venueid number(2), Equipstat number(1))
CREATE TABLE Incourse	(Instructid char(3), Courseid char(10), Nooftimes number(3), Courseowner char(1), Writer char(1))
CREATE TABLE Instime	(Instructid char(3), Commno number(6), Weekno number(3))

14.3   Compiled program suite

Three subprograms have been identified:

"Schedule Trainer for Course"
"OK"
"Not OK"

The Apple Macintosh, HyperCard and Oracle SQL code for these subprograms is given below.

14.3.1   *"Schedule Trainer for Course"*

```
on mouseUp
    global Minsid, Mweekno, Mnoofwks, Mcourseid, Mseqno
    set the cursor to 4
    execsql "open database logon roger identified by roger"
    execsql "select courseid, seqno, weekno into :Mcourseid:," && ¬
    ":Mseqno:, :Mweekno:" && ¬
```

```
"from offering where instructid = ' '"
execsql "get next row"
if the result = 24 then
    put "No instructorless courses" into card field "info"
    put −100 into Minsid
    execsql "close database"
    exit mouseUp
end if
put Mcourseid into card field "Coursename"
execsql "select noofwks into :Mnoofwks:," && ¬
"from course where courseid = :Mcourseid:"
execsql "get next row"
execsql "open cursor :cursorOne:"
execsql "open cursor :cursorTwo:"
execsql "select instructid into :Minsid: from incourse" && ¬
"where courseid = :Mcourseid: with cursor :cursorOne:"
put Mweekno into Mweek
repeat
    put −99 into Minsid
    execsql "get next row of :cursorOne:"
    if the result = 24 then
        exit repeat
    end if
    execsql "select commno into :Mcommno: from instime" && ¬
    "where weekno = :Mweek: AND instructid = :Minsid:" && ¬
    "with cursor :cursorTwo:"
    repeat
        put −99 into Mcommno
        execsql "get next row of :cursorTwo:"
        if Mcommno > 0 then
            exit repeat
        end if
        add 1 to Mweek
        if Mweek > Mweekno + Mnoofwks −1 then
            exit repeat
        end if
    end repeat
    if Mcommno < 0 then
        exit repeat
    end if
end repeat
if Minsid = −99 then
    put "No instructor free" into card field "info"
else
    put Minsid into card field 1
```

```
        end if
      execsql "close database"
    end mouseUp
```

14.3.2   "OK"

```
on mouseUp
  global Minsid, Mweekno, Mnoofwks, Mcourseid, Mseqno
  set the cursor to 4
  if Minsid = −100 then
    put " " into card field "info"
    exit mouseUp
  end if
  if Minsid = −99 then
    put "???" into Minsid
  end if
  execsql "open database logon roger identified by roger"
  — select offering based on Courseid and seqno and
  — update the instructid field
  execsql "update offering set instructid = :Minsid:" && ¬
  "where courseid = :Mcourseid: AND seqno = :Mseqno:"
  if Minsid = "???" then
    put −100 into Minsid
    execsql "close database"
    put " " into card field "Coursename"
    put " " into card field 1
    put " " into card field "info"
    exit mouseUp
  end if
  execsql "select commno into :Mcommno: from instime" && ¬
  "where instructid = 'xxx'"
  execsql "get next row"
  repeat
    add 1 to Mcommno
    execsql "insert into instime" && ¬
    "values (:Minsid:, :Mcommno:, :Mweekno:)"
    add 1 to Mweekno
    subtract 1 from Mnoofwks
    if Mnoofwks = 0 then
      exit repeat
    end if
  end repeat
  execsql "update instime set commno = :Mcommno:" && ¬
  "where instructid = 'xxx'"
  execsql "close database"
```

```
    put − 100 into Minsid
    put " " into card field "Coursename"
    put " " into card field 1
    put " " into card field "info"
end mouseUp
```

14.3.3 "Not OK"

```
on mouseUp
    global Minsid, Mweekno, Mnoofwks, Mcourseid, Mseqno
    set the cursor to 4
    if Minsid = − 100 then
        exit mouseUp
    end if
    execsql "open database logon roger identified by roger"
    execsql "update offering set instructid = '???'" && ¬
    "where courseid = :Mcourseid: AND seqno = :Mseqno:"
    execsql "close database"
    put − 100 into Minsid
    put " " into card field "Coursename"
    put " " into card field 1
    put " " into card field "info"
end mouseUp
```

14.4 Produce tests

The special HyperCard card entitled 'Testing RCT', described in the Design Phase
as modified design screen 14411, is set up. The test data consist of:

1. Five trainers (to be stored in INSTRUCT).
2. Each of the trainers can give any of the courses (INCOURSE).
3. Four different types of course (COURSE).
4. Eight course offerings which require a trainer scheduled for them
   (OFFERING).

The schedule for the course offering is summarised in Figure 14.2.
   The full HyperCard script and Oracle SQL code for setting up the databases
and the test data are given below:

```
on mouseUp
    set the cursor to 4
    execsql "open database logon roger identified by roger"
    execsql "drop table incourse"
    execsql "drop table instruct"
    execsql "drop table course"
    execsql "drop table instime"
```

Figure 14.2
Schedule for courses
for testing purposes.

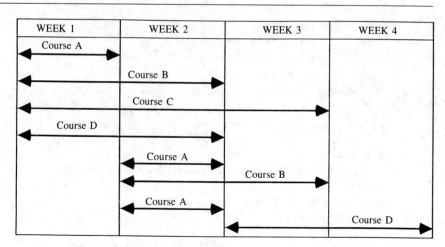

```
execsql "drop table offering"
execsql "create table course" && ¬
   "(courseid char(10)," && ¬
   "coursenm char(40)," && ¬
   "noofwks number(1)," && ¬
   "noofdays number(2)," && ¬
   "maxplaces number(2)," && ¬
   "equipnotes char(70))"
execsql "insert into course" && ¬
   "values('CourseA', 'RCT — Business Analysis', 1, 2, 15," && ¬
   "'special overheads')"
execsql "insert into course" && ¬
   "values('CourseB', 'RCT — Data Analysis', 2, 3, 15, ' ')"
execsql "insert into course" && ¬
   "values('CourseC', 'RCT — Activity Analysis', 3, 5, 15 ' ')"
execsql "insert into course" && ¬
   "values('CourseD', 'RCT — Analysis for Users', 2, 5, 7,
'Posh rooms')"
execsql "create table incourse" && ¬
   "(instructid char(3), courseid char(10), nooftimes number(3)," && ¬
   "courseowner char(1), writer char(1))"
execsql "insert into incourse" && ¬
   "values('sam', 'CourseA', 1, 'N', 'N')"
execsql "insert into incourse" && ¬
   "values('sam', 'CourseB', 1, 'N', 'N')"
execsql "insert into incourse" && ¬
   "values('sam', 'CourseC', 1, 'N', 'N')"
execsql "insert into incourse" && ¬
   "values('sam', 'CourseD', 1, 'N', 'N')"
execsql "insert into incourse" && ¬
```

"values('rak', 'CourseA', 1, 'N', 'N')"
execsql "insert into incourse" && ¬
"values('rak', 'CourseB', 1, 'N', 'N')"
execsql "insert into incourse" && ¬
"values('rak', 'CourseC', 1, 'N', 'N')"
execsql "insert into incourse" && ¬
"values('rak', 'CourseD', 1, 'N', 'N')"
execsql "insert into incourse" && ¬
"values('tp', 'CourseA', 1, 'N', 'N')"
execsql "insert into incourse" && ¬
"values('tp', 'CourseB', 1, 'N', 'N')"
execsql "insert into incourse" && ¬
"values('tp', 'CourseC', 1, 'N', 'N')"
execsql "insert into incourse" && ¬
"values('tp', 'CourseD', 1, 'N', 'N')"
execsql "insert into incourse" && ¬
"values('lt', 'CourseA', 1, 'N', 'N')"
execsql "insert into incourse" && ¬
"values('lt', 'CourseB', 1, 'N', 'N')"
execsql "insert into incourse" && ¬
"values('lt', 'CourseC', 1, 'N', 'N')"
execsql "insert into incourse" && ¬
"values('lt', 'CourseD', 1, 'N', 'N')"
execsql "insert into incourse" && ¬
"values('rjh', 'CourseA', 1, 'N', 'N')"
execsql "insert into incourse" && ¬
"values('rjh', 'CourseB', 1, 'N', 'N')"
execsql "insert into incourse" && ¬
"values('rjh', 'CourseC', 1, 'N', 'N')"
execsql "insert into incourse" && ¬
"values('rjh', 'CourseD', 1, 'N', 'N')"
execsql "create table instruct" && ¬
"(instructid char(3), surname char(15), firstnm char(8))"
execsql "insert into instruct" && ¬
"values('sam', 'Manspond', 'Steve')"
execsql "insert into instruct" && ¬
"values('rak', 'Kidney', 'Robin')"
execsql "insert into instruct" && ¬
"values('tp', 'Perry', 'Theo')"
execsql "insert into instruct" && ¬
"values('lt', 'Thell', 'Leslie')"
execsql "insert into instruct" && ¬
"values('rjh', 'Hipperson', 'Roger')"
execsql "create table instime" && ¬
"(instructid char(3), commno number(6), weekno number(3))"

```
execsql "create table offering" && ¬
    "(courseid char(10)," && ¬
    "seqno number(3)," && ¬
    "weekno number(3)," && ¬
    "startday number(1)," && ¬
    "status number(1)," && ¬
    "geocode number(2)," && ¬
    "instructid char(3)," && ¬
    "venueid number(2)," && ¬
    "equipstat number(1))"
execsql "insert into offering" && ¬
    "values('CourseA', 1, 1, 2, 0, 1, ' ', 0, 0)"
execsql "insert into offering" && ¬
    "values('CourseB', 1, 1, 2, 0, 1, ' ', 0, 0)"
execsql "insert into offering" && ¬
    "values('CourseC', 1, 1, 2, 0, 1, ' ', 0, 0)"
execsql "insert into offering" && ¬
    "values('CourseD', 1, 1, 2, 0, 1, ' ', 0, 0)"
execsql "insert into offering" && ¬
    "values('CourseA', 2, 2, 2, 0, 1, ' ', 0, 0)"
execsql "insert into offering" && ¬
    "values('CourseB', 2, 2, 2, 0, 1, ' ', 0, 0)"
execsql "insert into offering" && ¬
    "values('CourseA', 3, 2, 2, 0, 1, ' ', 0, 0)"
execsql "insert into offering" && ¬
    "values('CourseD', 2, 3, 2, 0, 1, ' ', 0, 0)"
execsql "insert into instime" && ¬
    "values('xxx', 0, 0)"
execsql "close database"
end mouseUp
```

One additional piece of data is stored, namely a dummy record in INSTIME giving the start commitment number of 0 against a dummy instructor ' '. At the start of the script, any existing tables are deleted, giving a new start before each test.

Test procedure

The test procedure is as follows:

1. Invoke "Set up data for test".
2. Invoke "Schedule Trainer for Course", and each time complete the transaction by invoking "OK".
3. The test is complete when the message "No instructorless courses" appears.

The expected results are shown in Table 14.1.

Table 14.1

Invoke button	Contents of "Course Name"	Contents of "Available Instructor"	Contents of reply field	Pass/fail	Comments
Set up data	blank	blank	blank		
Schedule trainer	CourseA	sam	blank		
OK	blank	blank	blank		
Schedule trainer	CourseB	rak	blank		
OK	blank	blank	blank		
Schedule trainer	CourseC	tp	blank		
OK	blank	blank	blank		
Schedule trainer	CourseD	lt	blank		
OK	blank	blank	blank		
Schedule trainer	CourseA	sam	blank		
OK	blank	blank	blank		
Schedule trainer	CourseB	rjh	blank		
OK	blank	blank	blank		
Schedule trainer	CourseA	blank	"No instructor free"		
OK	blank	blank	blank		
Schedule trainer	CourseD	sam	blank		
OK	blank	blank	blank		
Schedule trainer	blank	blank	"No instructorless courses"		

14.5   Run tests

The tests were run over the period 17−20 July 1990. The results are shown in Table 14.2. Further comments:

Table 14.2

Invoke button	Contents of "Course Name"	Contents of "Available Instructor"	Contents of reply field	Pass/fail	Comments
Set up data	blank	blank	blank	√	
Schedule trainer	CourseA	sam	blank	√	
OK	blank	blank	blank	√	
Schedule trainer	CourseB	rak	blank	√	
OK	blank	blank	blank	√	
Schedule trainer	CourseC	tp	blank	√	
OK	blank	blank	blank	√	
Schedule trainer	CourseD	lt	blank	√	
OK	blank	blank	blank	√	
Schedule trainer	CourseA	sam	blank	√	
OK	blank	blank	blank	√	
Schedule trainer	CourseB	rjh	blank	√	
OK	blank	blank	blank	√	
Schedule trainer	CourseA	blank	"No instructor free"	√	
OK	blank	blank	blank	√	
Schedule trainer	CourseD	sam	blank	√	
OK	blank	blank	blank	√	
Schedule trainer	blank	blank	"No instructorless courses"	√	

1. "No instructors free" message and "No instructorless courses" look very similar.
2. The message field is not very well positioned.
3. Not very happy that I couldn't get at the version number of the procedure.
4. I note that the "NOT OK" button is not tested.

Arthur Payne
Chief Tester
27/08/90

14.6   Quality review

14.6.1   Quality Review Team

The quality team under Mr N.O. Gotos reviewed the HyperCard scripts over the period 17— 20 July.

14.6.2   Comments

The following comments were made and reluctantly agreed by the Project Team:

1. None of the scripts has a standard header listing author, date of last change or version number.
   ACTION A: These must be added before acceptance.
2. There are far too few comments, and those that do exist are not meaningful enough.
   ACTION B: These must be added before acceptance.
3. There is no use of the SQL "JOIN" commands, which would potentially speed up the program.
   REPLY: This was to make the programs easier to understand following design into implementation.
   ACTION C: Reply accepted — no further action.
4. The technique of opening and closing the database between each transaction is very wasteful and may not be technically correct if the system is to be multi-user.
   REPLY: Good point — the design needs a review. We also believe that the user may request a design enhancement whereby all available instructors for any one course offering are displayed, allowing the user to select one. This will require substantial rework.
   ACTION D: Ensure that the requirement is correct and justify final design.

14.6.3   Review conclusion

1. This program suite is NOT of a suitable quality to be put forward for acceptance.
2. The Project Manager must make a decision on Action D.
3. The Quality Team wish to review the HyperCard scripts to ensure that Actions A and B have been suitably completed.

QUALITY UNACCEPTABLE

Mr N.O. Gotos
27/08/90

Bibliography

Introduction

This book is not an academic book: more a set of basic techniques as used in real life. The book has concentrated on the partnership between managers, users and analysts.

I have avoided making a long list of academic books, or even a list of all the books on my bookshelf. Instead, I have listed those books which have influenced me in some way, and into which I quite often delve for help. I have listed them under three headings:

1. *General.* Being general books on life that have some relevance to the role of understanding the business requirements before building a support system for them.
2. *IT in general.* Being IT books that I find have some relevance to the subject, but are not specific to analysis, design and/or implementation.
3. *Technical.* Being computing books which are well put together. They explain in much more detail the basic techniques that I have described. Usually, the techniques are explained in the context of a specific commercial method.

General

Thriving on Chaos, Tom Peters (Macmillan, 1988).
   Absolutely essential reading for everyone, whether or not you like the American style of Tom Peters.

The Renewal Factor, Robert H. Waterman (Bantam, 1989)
   Although not immediately obvious as relevant, this book is full of little gold nuggets. It is quite easy to read and very practical.

I am Right — You are Wrong, Edward de Bono (Viking, 1990)
   Written in Bono's style and full of 'obvious' bits of common sense. Tries to state that the world would be a better place if we got rid of conflict in our discussions. Sounds reasonable to me.

The Strategy of Indirect Approach, Liddell Hart (Faber and Faber, 1941)
   This book is all about the strategies of wars, with much relevance to the Second World War. However, the chapter entitled 'Concentrated Essence of Strategy' is as important to performing analysis as it is, presumably, to waging wars. In summary, Liddell Hart gives eight maxims for a successful strategy:
   1. Adjust your end to your means.
   2. Keep your object always in mind.
   3. Choose the line (or course) of least expectation.

4. Exploit the line of least resistance.
5. Take a line of operation which offers alternative objectives.
6. Ensure that both plan and dispositions are flexible — adaptable to circumstances.
7. Do not throw your weight into a stroke whilst your opponent is on guard.
8. Do not renew an attack along the same line (or in the same form) after it has once failed.

Information Technology in general

Peopleware, DeMarco and Lister (Dorset House/John Wiley, 1987)
This book is subtitled *Productive Projects and Teams*. It is full of excellent practical advice on how to run computer style projects, including having fun, and a section on quality.

Technical computing

SSADM Version 4 Reference Manual (NCC/Blackwell, 1990)
SSADM is the government preferred method in the UK. Whatever your views of SSADM, this manual is one of the most complete definitions of a method I have seen in the public domain and not associated directly with a supplier's computer support tool. Even if you have no intention of using SSADM, I thoroughly recommend reading through sections of this manual.

System Development, Michael Jackson (Prentice Hall, 1983)
An excellent book which describes the Michael Jackson way of doing things, which is somewhat different from most other techniques. The book is full of useful examples in using the techniques and covers quite a lot of the specifying and designing phases of the life-cycle.

Information Engineering (various volumes), James Martin (Savant, 1986)
This set of manuals describes the techniques of Information Engineering as at 1986. Information Engineering is a methodology used in the USA and in various northern European countries. Newer versions of the method have been Commercial-in-Confidence, held by the main Information Engineering players, namely Texas Instruments, Knowledgeware and James Martin Associates. It may be that, as this book is published, more up-to-date books on Information Engineering will make their way into the public domain. I have included Information Engineering in this book list mainly because the sum total of SSADM, JSD and Information Engineering covers just about every technique known in the commercial programming world.

Index

184

Selected figures

Key case study deliverables